SETS, LOGIC, AND AXIOMATIC THEORIES

Sets, Logic, and Axiomatic Theories

Second Edition

Robert R. Stoll

The Cleveland State University

W. H. Freeman and Company
SAN FRANCISCO

Library of Congress Cataloging in Publication Data

Stoll, Robert R
    Sets, logic, and axiomatic theories.

    Includes bibliographies.
    1.   Set theory.   2.   Logic, Symbolic and mathematical.
I.   Title.
QA248.3.S78   1974   511'.3   74–8932
ISBN 0–7167–0457–9

74008932

(14/2/94)

271593

To Benjamin

Contents

Preface to the Second Edition

ALTHOUGH UNDERGRADUATE MATHEMATICS curricula have changed since 1961, when the first edition of this book appeared, students who plan to study some abstract mathematics still need a knowledge of intuitive set theory as a working tool and will profit from an understanding of the notion of an axiomatic theory, as that concept is currently understood. Further, it is now conceded by some that aspects of mathematical logic are useful to such students. Recently, other consumers of these goods have emerged, for example, students of computer science.

This book has contributions to make in courses where one or more of the topics mentioned in its title are covered. Further, in its entirety it can bridge the gap between an undergraduate's initial conception of mathematics as a computational theory and the abstract nature of more advanced and more modern mathematics. Its skeleton is the same as that of the first edition, but the flesh is firmer and more extensive. I will elaborate on the latter assertion as I describe the coverage of the book. First, let me remark that I seriously considered placing Chapter 2 (Logic) ahead of Chapter 1 (Sets and Relations) on the ground that the sooner students gain exposure to the symbolism of logic and the construction of proofs, the better. That I did not reflects my continuing belief that naive set theory is easier than symbolic logic for a student beginning the study of abstract mathematics. Moreover, I feel that it provides fertile ground in which students can acquire exposure to, and gain experience in, the construction of informal proofs. It is at that point, I believe, that they are prepared to appreciate what symbolic logic with its formalism has to offer. However, an instructor who prefers to begin with Chapter 2 will encounter no difficulties with students who have had some exposure to the algebra of sets and the concept of a function.

Those topics in set theory that are essential for a course in modern algebra, topology, or analysis are covered in Chapter 1. New to this edition is a short section on the axiom of choice and a rather long section on proof and definition by induction. The latter includes an introductory account of the arithmetic of the system of natural numbers. Further, there is an extended bibliography, with specific references in the text and the footnotes to the titles listed in it. Some references in Chapter 1 are to original papers that played an important role in the development of set theory. When viewed in historical perspective, they prove to be exciting reading. Most references are to sources for further reading about topics that are not fully treated in the text or about details that have been omitted.

Chapter 2, which is devoted to mathematical logic, has been extensively revised. The exposition of the statement calculus, which is given in considerable detail, has been spruced up. First-order logic is barely more than outlined. However, since the outline follows the same pattern as that employed for the statement calculus, the momentum gained in the study of the statement calculus should make that of the predicate calculus intelligible. In this edition, function symbols have been incorporated into the presentation of first-order logic, and formal languages are discussed in some detail. Also, the exposition of the interpretation and validity of formulas of a formal language is now based on the concept of truth via satisfaction (instead of logical functions). Finally, the discussion of the notion of consequence for formal languages has been expanded slightly.

The first three sections of Chapter 3 discuss axiomatic theories as they occur in everyday mathematics. The rest of the chapter now develops the concepts of a formal axiomatic theory and of a first-order theory in some detail. This part of the chapter treats the syntax of first-order languages, thereby complementing the latter part of Chapter 2, where the semantics of such languages are treated. The section on metamathematics is now more up to date. An appendix on recursive functions has been added to make possible an understanding of the significance of Church's thesis.

Chapter 4 is concerned with the theory of Boolean algebras through the representation theorem. This is offered as a reward for those who have struggled through Chapters 1 and 2 and the first part of Chapter 3. Many of the concepts discussed there are put to use to obtain, in relatively few pages, a complete picture of the elementary part of a theory that has both historical and current interest.

My final product, this book, has profited greatly from criticisms and suggestions made by reviewers of an earlier draft. In particular I am most grateful to Professor Kurt Bing, who read that draft page-for-page and made numerous suggestions for improvements in both content and clarity.

January, 1974 Robert R. Stoll

SETS, LOGIC, AND AXIOMATIC THEORIES

CHAPTER ONE

Sets and Relations

THE THEORY OF SETS as a mathematical discipline originated with the German mathematician, G. Cantor (1845–1918). A complete account of its birth and childhood is out of the question here, since a considerable knowledge of mathematics is a prerequisite for its comprehension. Instead, we adopt the uneasy compromise of a brief sketch of these matters. If this proves unsatisfactory to the reader, nothing is lost; on the other hand, if it is at least partially understood, something may be gained.

Cantor's investigation of questions pertaining to trigonometric series and series of real numbers led him to recognize the need for a means of comparing the magnitude of infinite sets of numbers. To cope with this problem, he introduced the notion of the power (or size) of a set by defining two sets to have the *same* power if the members of one can be paired with those of the other. Since two finite sets can be paired if and only if they have the same number of members, the power of a finite set may be identified with a counting number. Thus the notion of power for infinite sets provides a generalization of everyday counting numbers. Cantor developed the theory, including an arithmetic, of these generalized (or transfinite) numbers and in so doing created a theory of sets. His accomplishments in this area are regarded as an outstanding example of mathematical creativity.*

* For a translation of two important papers by Cantor on transfinite numbers, together with a historical survey of his work, see Cantor (1915).

I refer several times in this chapter to original papers that had a great impact on the development of set theory. I do so in order to encourage you to look at these papers, for they

1

Cantor's insistence on dealing with the infinite as an actuality—he regarded infinite sets and transfinite numbers as being on a par with finite sets and counting numbers—was an innovation at that time. Since antiquity a majority of mathematicians had carefully avoided the introduction into their arguments of the *actual infinite* (that is, of sets containing an infinity of objects conceived as existing simultaneously, at least in thought). Since this attitude persisted until almost the end of the nineteenth century, Cantor's work was roundly criticized for supposedly encroaching on the domain of philosophers and violating principles of religion. However, as mathematicians discovered that some of Cantor's results had applications to analysis, attitudes began to change, and in the 1890's his ideas and results gained acceptance. By 1900, set theory had been accepted as an indispensable tool by a large segment of the mathematical community.

Another German mathematician, R. Dedekind (1831–1916), followed the work of Cantor with sustained interest from its beginning, since his thoughts on the concept of number had led him to set-theoretic matters. In his little book *Was sind und was sollen die Zahlen* (1888),* he showed how the concept of natural number can be derived from the basic notions of set theory. Since the systems of integers, of rational numbers, of real numbers, and of complex numbers can be developed, in succession, from the natural numbers with a certain amount of set theory [for such a development, see Stoll (1963)], it followed that the whole of classical mathematics could be based on set theory. For generations mathematicians had attempted to unify their discipline under a small number of basic notions. Thus, those concerned with foundational matters began to regard set theory not simply as a mathematical tool but also as a possible single source from which all known mathematics could be derived. (Subsequent developments have established that a carefully formulated theory of sets can indeed serve as a source of practically the whole of known mathematics.)

Ironically, around the turn of the century, at just about the time when Cantor's theory had gained general acceptance, several contradictions were derived within its superstructure. That these were not initially regarded as serious defects is indicated by the fact that they were called "paradoxes," defects which could be resolved once full understanding

---

have a freshness that is lost when their results are digested in a textbook. Further, reading them can help one realize that mathematics is a living subject, that mathematical theories do not come ready-made but are the results of groping and struggling. One source for many important papers is van Heijenoort (1967), in which all the papers have been translated into English and are accompanied by his illuminating comments.

* This is available in translation; see Dedekind (1901).

was acquired. Then B. Russell (1872–1970) discovered what became known as the Russell paradox (see §1.2) in 1901. Since it involves the most elementary aspects of set theory, it could not be ignored. The reaction of many mathematicians was that set theory should be provided with an axiomatic basis comparable to that which had been devised for Euclidean geometry (the axiomatic method is discussed in Chapter 3). The first axiomatization of set theory was given in 1908 by E. Zermelo (1871–1953).* Zermelo's system, with certain modifications by others, is still widely used [for a detailed account, see Stoll (1963)].

A formulation of set theory as an axiomatic theory would be too ambitious a goal for this book, in which the objective is instead to acquaint the reader with the basic set-theoretical notions and with how important mathematical concepts can be formulated in terms of these notions. Specifically, this chapter discusses, within the framework of set theory, four mathematical concepts: function, equivalence relation, ordering relation, and natural number. Sections 1.3 to 1.6 contain the necessary preliminaries, and Sections 1.1 and 1.2 describe the prerequisite principles of set theory, as used by Cantor. Why choose a starting point which is known to lead to contradictions? Because the important items in this chapter are independent of those features which characterize the Cantorian or "naive" approach to set theory. Indeed, any theory of sets, if it is to serve as a basis for mathematics, will include the principal definitions and theorems appearing in this chapter. Only the methods we employ to obtain some of these results are naive. No irreparable harm results from using such methods; they are standard tools in mathematics [see Cantor (1915), p. 85].

In this chapter we assume that the reader is familiar with the systems of integers, rational numbers, real numbers, and complex numbers, since knowledge in these areas enlarges the possibilities for constructing examples to assist the assimilation of definitions, theorems, and so on. We shall reserve the letters $\mathbb{Z}$, $\mathbb{Q}$, $\mathbb{R}$, and $\mathbb{C}$ for the set of integers, rational numbers, real numbers, and complex numbers, respectively and the symbols $\mathbb{Z}^+$, $\mathbb{Q}^+$, and $\mathbb{R}^+$ for the set of positive integers, positive rationals, and positive reals, respectively.

§1.1. Cantor's Concept of a Set

Let us discuss what the term "set" meant to Cantor. In the first of his major papers on the subject is the statement that a **set** S is any collection

---

* Zermelo's first paper may be found in van Heijenoort (1967).

of definite, distinguishable objects of our intuition or of our intellect to be conceived as a whole. The objects are called the **elements** or **members** of S. At best this is merely an intuitive description of the notion Cantor had in mind.* A distinguishing feature of the concept is that a collection of objects is to be regarded as a single entity (to be conceived as a whole). The transfer of attention from individual objects to collections of individual objects as entities is commonplace, as evidenced by the presence in our language of such words as "bunch," "covey," "pride," and "flock."

For the objects which may be allowed in a set, the phrase "objects of our intuition or of our intellect" gives considerable freedom. First, it gives complete liberty concerning the nature of the objects making up a set. Green apples, grains of sand, or prime numbers are admissible constituents of sets. However, for mathematical applications it is reasonable to choose as members such mathematical entities as points, lines, numbers, sets of numbers, and so on. Second, it permits the consideration of sets whose members cannot, for one reason or another, be explicitly exhibited. In this connection one is likely to think first of infinite sets, for which it is not even theoretically possible to collect the members as an assembled totality. The set of all prime numbers and the set of all points of the Euclidean plane having rational coordinates in a given coordinate system are examples of such sets.

On the other hand, there are finite sets which display the same degree of intangibility as any infinite set. To take an old example, begin with the premise that a typesetting machine with 10,000 characters (these would include the lower-case and capital letters of existing alphabets in various sizes of type, numerals, punctuation, and a blank character for spacing) would be adequate for printing in any language. (The *exact* size of the set of characters is not at issue; the reader may substitute for 10,000 any integer greater than 1.) Let it be agreed that by a "book" is meant a printed assemblage of 1,000,000 characters, including blank spaces. Thus a book may contain from 0 to 1,000,000 actual characters. Now consider the set of all books. Since there are 10,000 possibilities available for each of the 1,000,000 positions in a book, the total number of books is equal to $10,000^{1,000,000}$. This is a large (but finite!) number. In addition to books of gibberish, there would appear in the set all textbooks ever written or planned, all newspapers ever printed, all govern-

---

* If one does not resort to an axiomatic treatment, whereby sets are "defined implicitly" by virtue of assumptions made about them, it seems that no more than an intuitive description of such a fundamental concept is possible.

ment pamphlets, all train schedules, all logarithm tables ever computed, and so on, and so on. The magnitude eludes comprehension to the same degree as does that of an infinite set.

The remaining key words in Cantor's concept of a set are "distinguishable" and "definite." The intended meaning of the former, as he used it, was this: For any pair of objects qualified to appear as elements of a particular set, one must be able to determine whether they are different or the same. The attribute "definite" is interpreted as meaning that, if one is given a set and an object, one can determine whether the object is, or is not, a member of the set. The implication is that a set is completely determined by its members.

§1.2. The Basis of Intuitive Set Theory

According to Cantor, a set is made up of objects called members or elements (we shall use both terms synonymously). The assumption that, if one is presented with a specific object and a specific set, one can determine whether or not that object is a member of that set means this: If the first blank in "_____ is a member of _____" is filled in with the name of an object, and the second with the name of a set, the resulting sentence can be classified as true or false. Thus, the notion of membership is a relation between objects and sets. We shall symbolize this relation by $\in$ and write

$$x \in A$$

if the object x is a member of the set A. If x is not a member of A, we shall write

$$x \notin A.$$

Further,

$$x_1, x_2, \ldots, x_n \in A$$

will be used as an abbreviation for "$x_1 \in A$ and $x_2 \in A$ and . . . and $x_n \in A$."

In terms of the membership relation, Cantor's assumption that a set is determined by its members may be stated in the following form.

The intuitive principle of extension. *Two sets are equal iff (if and only if) they have the same members.*

The equality of two sets X and Y will be denoted by

$$X = Y,$$

and the inequality of X and Y by

$$X \neq Y.$$

Remark

A comment on the meaning of the concept of *identity* or *equality*, as it is used in mathematics, may be helpful. To each of "x is identical to y," "x is the same as y," and "x equals y" is attributed the same meaning, and each is symbolized by $x = y$. The basic assumption about equality, which dates back to G. Leibniz (1646–1716), is that $x = y$ just in the case that x and y share exactly the same properties. For set theory, where the membership relation is the sole primitive relation, this assumption takes the form

(1) $x = y$ iff for all sets A, $x \in A$ iff $y \in A$.

In the light of (1), we may weaken the principle of extension (a further assumption about equality) to

(2)   if for all a, $a \in X$ iff $a \in Y$, then $X = Y$,

since from (1) and (2) the converse of (2) can be derived. This converse is

(3)   if $X = Y$, then, for all a, $a \in X$ iff $a \in Y$.

From (1) follows

$$x = y \text{ and } x \in A \text{ imply } y \in A,$$

which asserts a substitutivity property of equality with respect to the first argument of $\in$. From (3) follows

$$X = Y \text{ and } a \in X \text{ imply } a \in Y,$$

which asserts a corresponding property of equality with respect to the second argument of $\in$.

It should be understood that the principle of extension is a nontrivial assumption about the membership relation. In general, a proof of the equality of two specified sets A and B is in two parts: one part demonstrates that if $x \in A$, then $x \in B$; the other demonstrates that if $x \in B$, then $x \in A$. An example of such a proof is given below.

That (uniquely determined) set whose members are the objects $x_1, x_2, \ldots, x_n$ will be written

$$\{x_1, x_2, \ldots, x_n\}.$$

In particular, $\{x\}$, a so-called **unit set** or a **singleton,** is the set whose sole member is x.

Examples

1.2.1. Let us prove that the set A of all positive even integers is equal to the set B of positive integers which are expressible as the sum of two positive odd integers. First we assume that $x \in A$ and deduce that $x \in B$. If $x \in A$, then

$x = 2m$, and hence $x = (2m - 1) + 1$, which means that $x \in B$. Next, we assume that $x \in B$ and deduce that $x \in A$. If $x \in B$, then $x = (2p - 1) + (2q - 1)$, and hence $x = 2(p + q - 1)$, which implies that $x \in A$. Thus, we have proved that A and B have the same members.

1.2.2. $\{2,4,6\}$ is the set consisting of the first three positive even integers. Since $\{2,4,6\}$ and $\{2,6,4\}$ have the same members, they are equal sets. Moreover, $\{2,4,6\} = \{2,4,4,6\}$ for the same reason.

1.2.3. The members of a set may themselves be sets. For instance, the National Football League is a set of football teams, each of which, in turn, is a set of individuals. Again, $\{\{1,3\}, \{2,4\}, \{5,6\}\}$ is a set with three members, namely, $\{1,3\}$, $\{2,4\}$ and $\{5,6\}$. The sets $\{\{1,2\}, \{2,3\}\}$ and $\{1,2,3\}$ are unequal, since the former has $\{1,2\}$ and $\{2,3\}$ as members, and the latter has 1, 2, and 3 as members.

1.2.4. The sets $\{\{1,2\}\}$ and $\{1,2\}$ are unequal, since the former, a unit set, has $\{1,2\}$ as its sole member and the latter has 1 and 2 as its members. This is an illustration of the general remark that an object and the set whose sole member is that object are distinct from each other.

We digress briefly to comment on the alphabets which we shall employ in discussing set theory. Usually, lower-case italic English letters will denote elements, and, for the time being, capital italic letters will denote sets which contain them. Later, lower-case Greek letters will be introduced for a certain type of set. If the members of a set are themselves sets, and if this is noteworthy in the discussion, capital script letters will be used for the containing set, and it will be called a **collection of sets.** For example, we might have occasion to discuss the collection $\mathcal{F}$ of all finite sets A of integers x. As a rule of thumb, the level of a set within a hierarchy of sets under consideration is suggested by the size and gaudiness of the letter employed to denote it.

Although the brace notation is practical for explicitly defining sets made up of a few elements, it is too unwieldly for defining sets having a large, finite number of elements and useless for infinite sets (sets having infinitely many elements). How can sets with a large number of elements be described? In this connection one instinctively tends to differentiate between finite and infinite sets on the grounds that a finite set can be realized as an assembled totality whereas an infinite set cannot. However, a large finite set (for example, the set of books described in §1.1) is as incapable of comprehension as is any infinite set. From such examples one must conclude that the problem of how to describe efficiently a large finite set and the problem of how to describe an infinite set are, for all practical purposes, one and the same.

A method that is commonly used consists of stating a property which is to be true for (or a condition which is to be satisfied by) those and only those objects which compose the set. The property assumes the role of a "defining property" for the set: an object is a member of the set iff it has the stated property. Plane analytic geometry abounds in examples of sets (of points) defined in this way. For instance, recall the definition of a parabola as the set of all points which are equidistant from a given line and a given point not on that line. In what follows, the notion of property, as we shall use it, is sharpened to some extent; a precise definition is possible only in terms of the concepts discussed in Chapter 2, beginning with §2.6.

Let us understand by a **statement** a declarative sentence which is either true or false. Now the expression

$$5 \text{ divides } x$$

is not a statement, for it is neither true nor false. Rather, it expresses a property of x. When we substitute for x the name of an appropriate object, we obtain a statement. The situation is conveniently described in terms of the notion of function as explained in elementary mathematics. The property of x expressed by "5 divides x" is a function of (the variable) x. This variable ranges over some set (the domain of the variable), $\mathbb{Z}$, let us say. To each member of this domain the function correlates a statement; that is, when x takes a member of $\mathbb{Z}$ as value, the property takes a statement as a corresponding value. Thus, the status of "5 divides x" is that it determines a statement function of one variable. To achieve harmony with definitions introduced in §2.6 we shall call this function a 1-*place predicate* or a *predicate in x*. This example illustrates our description of a *predicate in x* as a function of x which takes a statement as value upon the replacement of (each occurrence of) x by the name of an appropriate object. A recipe ("5 divides x" in our example) for computing values of a predicate in x is called a *predicate form in x*; a predicate form in x is said to *express* a predicate in x. We shall denote predicate forms by

$$P(x), \ Q(x), \text{ and so on.}$$

Our objective, that of defining sets in terms of properties, is achieved by accepting the following principle.

The intuitive principle of abstraction. *A predicate in x defines a set A by the convention that if $P(x)$ expresses the predicate then the members of A are exactly those objects a such that $P(a)$, the result of replacing each occurrence of x in $P(x)$ by an occurrence of a, is a true statement.*

Because sets having the same members are equal, the predicate expressed by $P(x)$ determines exactly one set, which we shall denote by

$$\{x|P(x)\},$$

read as "the set of all x such that $P(x)$." Thus,

$$a \in \{x|P(x)\} \text{ iff } P(a) \text{ is a true statement.}$$

If we think of $P(x)$ as expressing a property of x, we may summarize the principle of abstraction with the assertion that every property determines a set.

If two predicate forms $P(x)$ and $Q(x)$ express the same predicate, then $\{x|P(x)\} = \{x|Q(x)\}$ by an application of the principle of extension. For example,

$$\{x|x \in A \text{ and } x \in B\} = \{x|x \in B \text{ and } x \in A\}$$

and

$$\{x|x \in \mathbb{Z}^+ \text{ and } x < 5\} = \{x|x \in \mathbb{Z}^+ \text{ and } (x+1)^2 \leq 29\}.$$

An expression which contains variables in addition to x may be a predicate form in x. For example, each of

$$x \in \mathbb{Z} \text{ and, for some } y \in \mathbb{Z}, x = 5y \text{ or } x = 5y + 1$$

and

$$x \in \mathbb{Z} \text{ and, for all } y \in \mathbb{Z}^+, x < y$$

are predicate forms in x because the occurrences of the variable y are "tied down" (the technical term is "bound") by the presence of the term "for some" in the first expression and "for all" in the second. In contrast, in each expression, no occurrence of x is preceded by either of these terms, and on this account x is said to be "free" in each. In general, the notation "$P(x)$" is intended to indicate that x is free in the expression denoted by $P(x)$. If y is a variable which does not occur in $P(x)$, then $P(y)$ expresses the same predicate as $P(x)$ and so $\{x|P(x)\} = \{y|P(y)\}$. However, if y occurs bound in $P(x)$, then $P(x)$ and $P(y)$ may express different predicates and hence determine different sets. For example,

$$\{x|x \in \mathbb{Z} \text{ and for some } y \in \mathbb{Z}, y > x\}$$
$$\neq \{y|y \in \mathbb{Z} \text{ and for some } y \in \mathbb{Z}, y > y\}.$$

Further details of the treatment of variables in such formulas are offered in §2.8.

Examples

1.2.5. The following are examples of easily recognized sets defined by properties.

(a) $\{x|x$ is an integer greater than 1 and having no divisors less than or equal to $x^{1/2}\}$.

(b) $\{x|x$ is a positive integer less than 9$\}$.

(c) $\{x|x$ is a line of slope 3 in a coordinate plane$\}$.

(d) $\{x|x$ is a continuous function on the closed interval from 0 to 1$\}$.

1.2.6. $\{x|x = x_1$ or $x = x_2$ or . . . or $x = x_n\}$ is the set we earlier agreed to denote by $\{x_1, x_2, \ldots, x_n\}$.

1.2.7. In some cases our language makes possible, by way of a property, a briefer definition of a finite set than can be achieved by an enumeration of the elements. For example, it is shorter to define a particular set of 100 people by the property "x is a senator" than by enumerating names of the members.

1.2.8. If A is a set, then $x \in A$ is a predicate form and may be used as a defining property of a set. Since $y \in \{x|x \in A\}$ iff $y \in A$, we have

$$A = \{x|x \in A\},$$

by virtue of the principle of extension.

Various modifications of the basic brace notation for sets are used. For example, it is customary to write

$$\{x \in A|P(x)\}$$

instead of $\{x|x \in A$ and $P(x)\}$ for the set of all objects which are both members of A and have the property expressed by $P(x)$. An alternative description of this set is "all members of A which have the property expressed by $P(x)$," and it is this description that the new notation emphasizes. As illustrations, $\{x \in \mathbb{R}|0 \leq x \leq 1\}$ denotes the set of all real numbers between 0 and 1 inclusive, and $\{x \in \mathbb{Q}^+|x^2 < 2\}$ denotes the set of all positive rationals whose square is less than 2.

If $P(x)$ expresses a property and f is a function, then

$$\{f(x)|P(x)\}$$

will be used to denote the set of all y for which there is an x such that x has the property expressed by $P(x)$ and $y = f(x)$. For example, instead of writing

$$\{y|\text{ there is an } x \text{ such that } x \text{ is an integer and } y = 2x\},$$

we shall write

$$\{2x|x \in \mathbb{Z}\}.$$

Again, $\{x^2|x \in \mathbb{Z}\}$ denotes the set of squares of integers. Such notations have natural extensions; in general, one's intuition is an adequate guide for interpreting examples. For instance, in a coordinate plane, where the points are identified by the members of the set $\mathbb{R}^2$ of all ordered pairs $\langle x, y \rangle^*$ of real numbers x and y, it is reasonable to interpret $\{\langle x, y \rangle \in \mathbb{R}^2 | y = 2x\}$ as the line through the origin having slope 2.

Cantorian set theory is founded on the principles of extension and abstraction, and a third principle, the axiom of choice (which Cantor employed without explicit mention), which is formulated in §1.10. A development of set theory based on these assumptions illustrates the axiomatic method as understood by Euclid. In this older sense, the objects of study are assumed to be known and the axioms merely express properties of the objects which are taken as evident from their construction. In contrast, in a development of set theory within the framework of modern axiomatics (see Chapter 3), axioms are stated first (in terms of the predicates "is equal to" and "is a member of") and serve to introduce or "define implicitly" the objects of study (sets).

We have already mentioned that contradictions can be derived within intuitive set theory. The source of trouble is the unrestricted use of the principle of abstraction. This principle appears to have been first formulated explicitly by G. Frege (1848–1925); he derived it as a consequence of his Basic Law V in Frege (1964). In 1901 B. Russell discovered that a contradiction could be derived from this principle. It stems from the set R defined by the predicate form

$$x \text{ is a set and } x \notin x,$$

and is derived as follows. First we establish that $R \notin R$. For proof, assume that $R \in R$. From the predicate form defining R it follows that $R \notin R$. Thus $R \in R$ (by assumption) and $R \notin R$ (by proof). From this contradiction it follows that $R \notin R$ by a law of logic (see §2.5). Since R is a set and $R \notin R$, we infer that R possesses the property required for membership in itself; that is, $R \in R$. The conclusion that $R \notin R$ and $R \in R$ is the **Russell contradiction** (or paradox).†

Exercises

1.2.1. Explain why $2 \in \{1,2,3\}$.

1.2.2. Is $\{1,2\} \in \{\{1,2,3\}, \{1,3\}, 1, 2\}$? Justify your answer.

1.2.3. Try to devise a set which is a member of itself.

---

* Here we are using a notation which will be discussed in detail later.

† Russell derived the Russell paradox by applying Cantor's proof that there is no greatest cardinal number to the universal class. He communicated the paradox to Frege in 1902. His letter and Frege's response can be read in van Heijenoort (1967). Frege received Russell's letter while the second volume of his *Grundgesetze der Arithmetik* was in press. He stated his reaction in a famous appendix to this second volume published in 1903. It appears as Appendix II in the translation of selected parts of Frege's two volumes by M. Furth; see Frege (1964).

1.2.4. Give an example of sets A, B, and C such that $A \in B$, $B \in C$, and not $A \in C$.

1.2.5. Describe in prose each of the following sets.

(a) $\{x \in \mathbb{Z} | x$ is divisible by 2 and x is divisible by 3$\}$.

(b) $\{x | x \in A$ and $x \in B\}$.

(c) $\{x | x \in A$ or $x \in B\}$.

(d) $\{x \in \mathbb{Z}^+ | x \in \{x \in \mathbb{Z} |$ for some integer y, $x = 2y\}$ and $x \in \{x \in \mathbb{Z} |$ for some integer y, $x = 3y\}\}$.

(e) $\{x^2 | x$ is a prime$\}$.

(f) $\{a/b \in \mathbb{Q} | a + b = 1$ and $a, b \in \mathbb{Q}\}$.

(g) $\{\langle x,y \rangle \in \mathbb{R}^2 | x^2 + y^2 = 1\}$.

(h) $\{\langle x,y \rangle \in \mathbb{R}^2 | y = 2x$ and $y = 3x\}$.

1.2.6. Prove that if a, b, c, and d are any objects, not necessarily distinct from one another, then $\{\{a\}, \{a,b\}\} = \{\{c\}, \{c,d\}\}$ iff both $a = c$ and $b = d$.

§1.3. Inclusion

We now introduce two further relations for sets. If A and B are sets, then A is **included in** B, symbolized by $A \subseteq B$, iff each member of A is a member of B. In this event one also says that A is a **subset** of B. Further, we agree that B **includes** A, symbolized by $B \supseteq A$, is synonymous with A is included in B. Thus, $A \subseteq B$ and $B \supseteq A$ each means that, for all x, if $x \in A$, then $x \in B$. The set A is **properly included in** B, symbolized by $A \subset B$ (or, alternatively, A is a **proper subset** of B, and B **properly includes** A), iff $A \subseteq B$ and $A \neq B$. For example, the set of even integers is properly included in the set $\mathbb{Z}$ of integers, and the set $\mathbb{Q}$ of rational numbers properly includes $\mathbb{Z}$.

Among the basic properties of the inclusion relation are the following:

$$X \subseteq X;$$
$$X \subseteq Y \text{ and } Y \subseteq Z \text{ imply } X \subseteq Z;$$
$$X \subseteq Y \text{ and } Y \subseteq X \text{ imply } X = Y.$$

The second property is the basis of the notation

$$X \subseteq Y \subseteq Z \text{ for } X \subseteq Y \text{ and } Y \subseteq Z.$$

The last is the formulation, in terms of the inclusion relation, of the two steps in a proof of the equality of two sets. That is, to prove that $X = Y$, one proves that $X \subseteq Y$ and then that $Y \subseteq X$.

For the relation of proper inclusion, only the analogue of the second property above is valid. The proof that $X \subset Y$ and $Y \subset Z$ imply $X \subset Z$ is required in an exercise at the end of this section. There the reader will also find further properties of proper inclusion, as it relates to inclusion.

Since beginners tend to confuse the relations of membership and inclusion, we shall take every opportunity to point out distinctions. At this time we note that the analogues for membership of the first two of the above properties for inclusion are false. For example, if X is the set of prime numbers, then $X \notin X$. Again, although $1 \in \mathbb{Z}$ and $\mathbb{Z} \in \{\mathbb{Z}\}$, it is not the case that $1 \in \{\mathbb{Z}\}$, since $\mathbb{Z}$ is the sole member of $\{\mathbb{Z}\}$.

We turn now to a discussion of the subsets of a set, that is, the sets included in a set. This is our first example of an important procedure in set theory—the formation of new sets from an existing set. The principle of abstraction may be used to define subsets of a given set. Indeed, if $P(x)$ is a predicate form in x and A is a set, then the predicate form

$$x \in A \text{ and } P(x)$$

determines that subset of A which we have already agreed to write as $\{x \in A | P(x)\}$. If A is a set and we choose $P(x)$ to be $x \neq x$, the result is $\{x \in A | x \neq x\}$, and this set, clearly, has no elements. The principle of extension implies that there can be only one set with no elements. We call this set the **empty set*** and symbolize it by

$$\varnothing.$$

The empty set is a subset of every set. To establish this, it must be proved that if A is a set, then each member of $\varnothing$ is a member of A. Since $\varnothing$ has no members, the condition is automatically fulfilled. Although this reasoning is correct, it may not be satisfying. An alternative proof which might be more comforting is an indirect one. Assume that it is false that $\varnothing \subseteq A$. This can be the case only if there exists some member of $\varnothing$ which is not a member of A. But this is impossible, since $\varnothing$ has no members. Hence, $\varnothing \subseteq A$ is not false; that is, $\varnothing \subseteq A$.

Each set $A \neq \varnothing$ has at least two distinct subsets, A and $\varnothing$. Moreover, each member of A determines a subset of A; if $a \in A$, then $\{a\} \subseteq A$. There are occasions when one wishes to speak not of individual subsets of a set, but of the set of all subsets of that set. The set of all subsets of a set A is the **power set** of A, symbolized by

$$\mathcal{P}(A).$$

Thus, $\mathcal{P}(A)$ is an abbreviation for

$$\{B | B \subseteq A\}.$$

---

* The empty set can be defined more simply as $\{x | x \neq x\}$.

For instance, if $A = \{1,2,3\}$, then

$$\mathcal{P}(A) = \{A, \{1,2\}, \{1,3\}, \{2,3\}, \{1\}, \{2\}, \{3\}, \varnothing\}.$$

As another instance of the distinction between the membership and inclusion relations, we note that if $B \subseteq A$, then $B \in \mathcal{P}(A)$, and if $a \in A$, then $\{a\} \subseteq A$ and $\{a\} \in \mathcal{P}(A)$.

The name "power set of A" for the set of all subsets of A has its origin in the case where A is finite; then $\mathcal{P}(A)$ has 2^n members if A has n members. To prove this, consider the following scheme for describing a subset B of $A = \{a_1, \ldots, a_n\}$: a sequence of n 0's and 1's, where the first entry is 1 if $a_1 \in B$ and 0 if $a_1 \notin B$, where the second entry is 1 if $a_2 \in B$ and 0 if $a_2 \notin B$, and so on. Clearly, the subsets of A can be paired with the set of all such sequences of 0's and 1's; for example, if $n = 4$, then $\{a_1,a_3\}$ determines, and is determined by, the sequence 1010. Since the total number of such sequences is equal to $2 \cdot 2 \cdot \cdots \cdot 2 = 2^n$, the number of elements of $\mathcal{P}(A)$ is equal to 2^n.

Exercises

1.3.1. Prove each of the following, using any properties of numbers that may be needed.
   (a) $\{x \in \mathbb{Z}|$ for an integer y, $x = 6y\} = \{x \in \mathbb{Z}|$ for integers u and v, $x = 2u$ and $x = 3v\}$.
   (b) $\{x \in \mathbb{R}|$ for a real number y, $x = y^2\} = \{x \in \mathbb{R}|x \geq 0\}$.
   (c) $\{x \in \mathbb{Z}|$ for an integer y, $x = 6y\} \subseteq \{x \in \mathbb{Z}|$ for an integer y, $x = 2y\}$.

1.3.2. Prove each of the following for sets A, B, and C.
   (a) If $A \subseteq B$ and $B \subseteq C$, then $A \subseteq C$.
   (b) If $A \subseteq B$ and $B \subset C$, then $A \subset C$.
   (c) If $A \subset B$ and $B \subseteq C$, then $A \subset C$.
   (d) If $A \subset B$ and $B \subset C$, then $A \subset C$.

1.3.3. Give an example of sets A, B, C, D, and E which satisfy the following conditions simultaneously: $A \subset B$, $B \in C$, $C \subset D$, and $D \subset E$.

1.3.4. Which of the following are true for all sets A, B, and C?
   (a) If $A \nsubseteq B$ and $B \nsubseteq C$, then $A \nsubseteq C$.
   (b) If $A \neq B$ and $B \neq C$, then $A \neq C$.
   (c) If $A \in B$ and $B \nsubseteq C$, then $A \nsubseteq C$.
   (d) If $A \subset B$ and $B \subseteq C$, then $C \nsubseteq A$.
   (e) If $A \subseteq B$ and $B \in C$, then $A \notin C$.

1.3.5. Show that for every set A, $A \subseteq \varnothing$ iff $A = \varnothing$.

1.3.6. Let $A_1, A_2, \ldots, A_n$ be n sets. Show that
$$A_1 \subseteq A_2 \subseteq \ldots \subseteq A_n \subseteq A_1 \quad \text{iff} \quad A_1 = A_2 = \ldots = A_n.$$

1.3.7. Give several examples of a set X such that each element of X is a sub-set of X.

1.3.8. List the members of $\mathcal{P}(A)$ if $A = \{\{1,2\}, \{3\}, 1\}$.

1.3.9. For each positive integer n, give an example of a set A_n of n elements such that for each pair of elements of A_n, one member is an element of the other.

§1.4. Operations for Sets

We continue with our description of methods for generating new sets from existing sets by defining two methods for composing pairs of sets. These so-called operations for sets parallel, in certain respects, the familiar operations of addition and multiplication for integers. The **union** (**sum, join**) of the sets A and B, symbolized by $A \cup B$ and read "A union B" or "A cup B," is the set of all objects which are members of either A or B; that is,

$$A \cup B = \{x \mid x \in A \text{ or } x \in B\}.$$

Here the inclusive sense of the word "or" is intended. Thus, by defi-nition, $x \in A \cup B$ iff x is a member of at least one of A and B. For example,

$$\{1,2,3\} \cup \{1,3,4\} = \{1,2,3,4\}.$$

The **intersection** (**product, meet**) of the sets A and B, symbolized by $A \cap B$ and read "A intersection B" or "A cap B," is the set of all objects which are members of both A and B; that is,

$$A \cap B = \{x \mid x \in A \text{ and } x \in B\}.$$

Thus, by definition, $x \in A \cap B$ iff $x \in A$ and $x \in B$. For example,

$$\{1,2,3\} \cap \{1,3,4\} = \{1,3\}.$$

It is left as an exercise to prove that for every pair of sets A and B the following inclusions hold:

$$\varnothing \subseteq A \cap B \subseteq A \subseteq A \cup B.$$

Two sets A and B are **disjoint** iff $A \cap B = \varnothing$, and they **intersect** iff $A \cap B \neq \varnothing$. A collection of sets is a **disjoint collection** iff each distinct pair of its member sets is disjoint. A **partition** of a set X is a disjoint collection $\mathcal{C}$ of nonempty and distinct subsets of X such that each member of X is a member of some (and, hence, exactly one) mem-ber of $\mathcal{C}$. For example, $\{\{1,2\}, \{3\}, \{4,5\}\}$ is a partition of $\{1,2,3,4,5\}$.

A further procedure, that of complementation, for generating sets from existing sets employs a single set. The **absolute complement** of a set A, symbolized by $\overline{A}$, is $\{x \mid x \notin A\}$. The **relative complement** of A

with respect to a set X is $X \cap \overline{A}$; this is usually shortened to $X - A$, read "X minus A." Thus $X - A$ is an abbreviation for

$$\{x \in X | x \notin A\},$$

that is, the set of those members of X which are not members of A. The **symmetric difference** of sets A and B, symbolized by $A + B$, is defined as follows:

$$A + B = (A - B) \cup (B - A).$$

This operation is commutative, that is, $A + B = B + A$, and associative, that is, $(A + B) + C = A + (B + C)$. Further, $A + A = \varnothing$, and $A + \varnothing = A$. Proofs of these statements are left as exercises.

If all sets under consideration in a certain discussion are subsets of a set U, then U is called the **universal set** (for that discussion). As examples, in elementary number theory the universal set is $\mathbb{Z}$, and in plane analytic geometry the universal set is the set of all ordered pairs of real numbers. A graphic device known as a Venn diagram is used for assisting one's thinking on complex relations which may exist among subsets of a universal set U. A Venn diagram is a schematic representation of sets by sets of points: the universal set U is represented by the points within a rectangle, and a subset A of U is represented by the interior of a circle or some other simple region within the rectangle. The complement of A relative to U, which we may abbreviate to $\overline{A}$ without confusion, is the part of the rectangle outside the region representing A, as shown in Figure 1.1. If the subsets A and B of U are represented in this way, then $A \cap B$ and $A \cup B$ are represented by shaded regions, as in Figure 1.2 and Figure 1.3, respectively. Disjoint sets are represented by nonoverlapping regions, and inclusion is depicted by displaying one region lying entirely within another. These are the ingredients for constructing the Venn diagram of an expression compounded from several sets by means of union, intersection, complementation, and inclusion. A Venn diagram for the symmetric difference of sets A and B appears in Figure 1.4 and one for $A \cup (B \cap C)$ is shown in Figure 1.5. The principal applications of Venn diagrams are to prob-

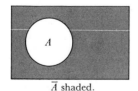

$\overline{A}$ shaded.

Figure 1.1.

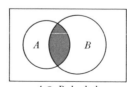

$A \cap B$ shaded.

Figure 1.2.

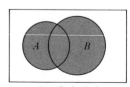

$A \cup B$ shaded.

Figure 1.3.

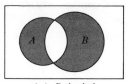

$A + B$ shaded.

Figure 1.4.

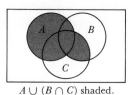

$A \cup (B \cap C)$ shaded.

Figure 1.5.

lems of simplifying a given complex expression and simplifying given sets of conditions among several subsets of a universe of discourse. Three simple examples of this sort appear below. In many cases such diagrams are inadequate, but they may be helpful in connection with the algebraic approach developed in the next section.

Examples

1.4.1. Suppose A and B are given sets such that $A - B = B - A = \emptyset$. Can the relation of A to B be expressed more simply? Since $A - B = \emptyset$ means $A \cap \bar{B} = \emptyset$, the regions representing A and $\bar{B}$ do not overlap (Figure 1.6). Clearly, $\bar{\bar{B}} = B$, so we conclude (Figure 1.7) that $A \subseteq B$. Conversely, if $A \subseteq B$, it is clear that $A - B - \emptyset$. We conclude that $A - B = \emptyset$ iff $A \subseteq B$. Interchanging A and B gives $B - A = \emptyset$ iff $B \subseteq A$. Thus the given relations hold between A and B iff $A \subseteq B$ and $B \subseteq A$ or, $A = B$.

1.4.2. Let us investigate the question of whether it is possible to find three subsets A, B, and C of U such that

$$C \neq \emptyset, \qquad A \cap B \neq \emptyset, \qquad A \cap C = \emptyset, \qquad (A \cap B) - C = \emptyset.$$

The second condition implies that A and B intersect and, therefore, incidentally that neither is empty. From Example 1.4.1 the fourth condition amounts to $A \cap B \subseteq C$, from which it follows that the first is superfluous. The associated Venn diagram indicates that A and C intersect; that is, the validity of the second and fourth conditions contradicts the third. Hence, there do not exist sets satisfying all of the conditions simultaneously.

1.4.3. Given that F, G, and L are subsets of U such that

$$F \subseteq G, \qquad G \cap L \subseteq F, \qquad L \cap F = \emptyset.$$

Is it possible to simplify this set of conditions? The Venn diagram (Figure 1.8)

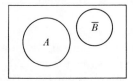

Figure 1.6.

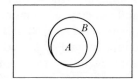

Figure 1.7.

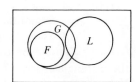

Figure 1.8.

represents only the first and third conditions. The second condition forces L and G to be disjoint; that is, $G \cap L = \emptyset$. On the other hand, if $F \subseteq G$ and $G \cap L = \emptyset$, then all given conditions hold. Thus $F \subseteq G$ and $G \cap L = \emptyset$ constitute a simplification of the given conditions.

Exercises

Note: Venn diagrams are *not* to be used in Exercises 1.4.1–1.4.8.

1.4.1. Prove that for all sets A and B, $\emptyset \subseteq A \cap B \subseteq A \cup B$.

1.4.2. Let $\mathbb{Z}$ be the universal set, and let

$$A = \{x \in \mathbb{Z} | \text{ for some positive integer } y, \ x = 2y\},$$
$$B = \{x \in \mathbb{Z} | \text{ for some positive integer } y, \ x = 2y - 1\},$$
$$C = \{x \in \mathbb{Z} | x < 10\}.$$

Describe $\bar{A}$, $A \cup B$, $\bar{C}$, $A - \bar{C}$, and $C - (A \cup B)$ either in prose or by a defining property.

1.4.3. Consider the following subsets of $\mathbb{Z}^+$, the set of positive integers:

$$A = \{x \in \mathbb{Z}^+ | \text{ for some integer } y, \ x = 2y\},$$
$$B = \{x \in \mathbb{Z}^+ | \text{ for some integer } y, \ x = 2y + 1\},$$
$$C = \{x \in \mathbb{Z}^+ | \text{ for some integer } y, \ x = 3y\}.$$

(a) Describe $A \cap C$, $B \cup C$, and $B - C$.

(b) Verify that $A \cap (B \cup C) = (A \cap B) \cup (A \cap C)$.

1.4.4. If A is any set, what is each of the following sets? $A \cap \emptyset$, $A \cup \emptyset$, $A - \emptyset$, $A - A$, $\emptyset - A$.

1.4.5. Determine $\emptyset \cap \{\emptyset\}$, $\{\emptyset\} \cap \{\emptyset\}$, $\{\emptyset, \{\emptyset\}\} - \emptyset$, $\{\emptyset, \{\emptyset\}\} - \{\emptyset\}$, $\{\emptyset, \{\emptyset\}\} - \{\{\emptyset\}\}$.

1.4.6. Suppose A and B are subsets of U. Show that, in each of (a), (b), and (c) below, if any of the relations stated holds, then each of the others holds.

(a) $A \subseteq B$, $\quad \bar{A} \supseteq \bar{B}$, $\quad A \cup B = B$, $\quad A \cap B = A$.

(b) $A \cap B = \emptyset$, $\quad A \subseteq \bar{B}$, $\quad B \subseteq \bar{A}$.

(c) $A \cup B = U$, $\quad \bar{A} \subseteq B$, $\quad \bar{B} \subseteq A$.

1.4.7. Prove that for all sets A, B, and C,

$$(A \cap B) \cup C = A \cap (B \cup C) \quad \text{iff} \quad C \subseteq A.$$

1.4.8. Prove that for all sets A, B, and C,

$$(A - B) - C = (A - C) - (B - C).$$

1.4.9. (a) Using a Venn diagram, show that symmetric difference is a commutative and associative operation.

(b) Show that for every set A, $A + A = \emptyset$ and $A + \emptyset = A$.

1.4.10. The Venn diagram for subsets A, B, and C of U, in general, divides the rectangle representing U into eight nonoverlapping regions (see Figure 1.5). Label each region with a combination of A, B, and C which represents exactly that region.

1.4.11. With the aid of a Venn diagram investigate the validity of each of the following inferences:

(a) If A, B, and C are subsets of U such that $A \cap B \subseteq \bar{C}$ and $A \cup C \subseteq B$, then $A \cap C = \varnothing$.

(b) If A, B, and C are subsets of U such that $A \subseteq \overline{B \cup C}$ and $B \subseteq \overline{A \cup C}$, then $B = \varnothing$.

§1.5. The Algebra of Sets

If we were to undertake the treatment of problems more complex than those examined above, we would feel the need for more systematic procedures for carrying out calculations with sets related by inclusion, union, intersection, and complementation. That is, what would be called for could appropriately be named "the algebra of sets"—a development of the basic properties of $\cup$, $\cap$, $^{-}$, and $\subseteq$ together with interrelations. As such, the algebra of sets is intended to be the set-theoretic analogue of the familiar algebra of the real numbers, which is concerned with properties of $+$, $\cdot$, and $\leq$ and their interrelations. The basic ingredients of the algebra of sets are various **identities**—equations which are true whatever the universal set U and no matter what particular subsets the letters (other than U and $\varnothing$) represent.

Our first result lists basic properties of union and intersection. For the sake of uniformity, all of these have been formulated for subsets of a universal set U. However, for some of the properties this is a purely artificial restriction, as an examination of the proofs will show.

THEOREM 1.5.1. For any subsets A, B, C of a set U the following equations are identities. Here $\bar{A}$ is an abbreviation for $U - A$.

1. $A \cup (B \cup C) = (A \cup B) \cup C.$

1'. $A \cap (B \cap C) = (A \cap B) \cap C.$

2. $A \cup B = B \cup A.$

2'. $A \cap B = B \cap A.$

3. $A \cup (B \cap C) = (A \cup B) \cap (A \cup C).$

3'. $A \cap (B \cup C) = (A \cap B) \cup (A \cap C).$

4. $A \cup \varnothing = A.$

4'. $A \cap U = A.$

5. $A \cup \bar{A} = U.$

5'. $A \cap \bar{A} = \varnothing.$

Proof. Each assertion can be verified by showing that the set on either side of the equality sign is included in the set on the other side. As an illustration we shall prove identity 3.

(a) Proof that $A \cup (B \cap C) \subseteq (A \cup B) \cap (A \cup C)$. Let $x \in A \cup (B \cap C)$. Then $x \in A$ or $x \in B \cap C$. If $x \in A$, then $x \in A \cup B$ and

$x \in A \cup C$, and hence x is a member of their intersection. If $x \in B \cap C$, then $x \in B$ and $x \in C$. Hence $x \in A \cup B$ and $x \in A \cup C$, so again x is a member of their intersection.

(b) Proof that $(A \cup B) \cap (A \cup C) \subseteq A \cup (B \cap C)$. Let $x \in (A \cup B) \cap (A \cup C)$. Then $x \in A \cup B$ and $x \in A \cup C$. Hence, $x \in A$, or $x \in B$ and $x \in C$. These imply that $x \in A \cup (B \cap C)$.

Identities 1 and $1'$ are referred to as the associative laws for union and intersection, respectively, and identities 2 and $2'$ as the commutative laws for these operations. Identities 3 and $3'$ are the distributive laws for union and intersection, respectively. The analogy of properties of union and intersection with properties of addition and multiplication, respectively, for numbers, is striking at this point. For instance, $3'$ corresponds precisely to the distributive law in arithmetic. That there are also striking differences is illustrated by 3, which has no analogue in arithmetic.

According to the associative law, identity 1, the two sets that can be formed with the operation of union from sets A, B, and C, in that order, are equal. We agree to denote this set by $A \cup B \cup C$. Then the associative law asserts that parentheses can be introduced in any order into this expression. Using induction (see Theorem 1.12.3), this result can be generalized to the following. The sets obtainable from given sets A_1, A_2, . . . , A_n, in that order, by use of the operation of union are all equal to one another. The set defined by A_1, A_2, . . . , A_n in this way will be written as

$$A_1 \cup A_2 \cup \cdots \cup A_n.$$

In view of identity $1'$ there is also a corresponding generalization for intersection. With these general associative laws on the record we can state the general commutative law: If $1'$, $2'$, . . . , n' are 1, 2, . . . , n in any order, then

$$A_1 \cup A_2 \cup \cdots \cup A_n = A_1' \cup A_2' \cup \cdots \cup A_n';$$

and the general distributive laws:

$$A \cup (B_1 \cap B_2 \cap \cdots \cap B_n)$$
$$= (A \cup B_1) \cap (A \cup B_2) \cap \cdots \cap (A \cup B_n),$$
$$A \cap (B_1 \cup B_2 \cup \cdots \cup B_n)$$
$$= (A \cap B_1) \cup (A \cap B_2) \cup \cdots \cup (A \cap B_n).$$

These can also be proved by induction.

Detailed proofs of the foregoing properties (precisely, equational identities) of union and intersection of sets need make no reference to the membership relation, for they may be deduced from the identities listed in Theorem 1.5.1. The same is true of the additional identities that appear in the next theorem. The discovery of such facts is at the roots of the "axiomatic approach" to the algebra of sets which is developed in Chapter 4. From this approach emerges the conclusion that *every* statement which is true for all algebras of sets is derivable from the identities 1 to 5 and 1' to 5'. This assertion is made precise and a proof is outlined in the last two paragraphs of §4.6.

These ten properties have another interesting consequence. In Theorem 1.5.1 they are paired in such a way that each member of a pair is obtainable from the other member by interchanging $\cup$ and $\cap$ and, simultaneously, $\varnothing$ and U. An equation, or an expression, or a statement within the framework of the algebra of sets obtained from another by interchanging $\cup$ and $\cap$ along with $\varnothing$ and U throughout is the **dual** of the original. We contend that the dual of any theorem expressible in terms of $\cup$, $\cap$, and $^{-}$, and which can be proved using only identities 1–5 and 1'–5', is also a theorem. Indeed, suppose that the proof of such a theorem is written as a sequence of steps and that opposite each step is placed the justification for it. By assumption, each justification is one of 1–5, one of 1'–5', or a premise of the theorem. Now replace the identity or relation in each step by its dual. Since 1–5 and 1'–5' contain with each its dual, and the dual of each premise of the original theorem is now a premise, the dual of each justification in the original proof is available to serve as a justification for a step in the new sequence which, therefore, constitutes a proof. The last line of the new sequence is, therefore, a theorem, the dual of the original theorem. Accepting the fact that every theorem of the algebra of sets is deducible from 1–5 and 1'–5', we then obtain the **principle of duality** for the algebra of sets: If T is any theorem expressed in terms of $\cup$, $\cap$, and $^{-}$, then the dual of T is also a theorem. This implies, for instance, that if the unprimed formulas in the next theorem are deduced solely from Theorem 1.5.1, then the primed formulas follow by duality. The reader should convince himself that all the assertions in Theorem 1.5.2 are true by using the definitions of $\cup$, $\cap$, and $^{-}$ in terms of the membership relation. Further, he might try to deduce some of them solely from Theorem 1.5.1, that is, without appealing in any way to the membership relation. Some demonstrations of this nature appear in the proof of Theorem 4.2.1.

THEOREM 1.5.2. For all subsets A and B of a set U, the following statements are valid. Here $\overline{A}$ is an abbreviation for $U - A$.

6. If, for all A, $A \cup B = A$,   6′. If, for all A, $A \cap B = A$,
   then $B = \varnothing$.   then $B = U$.
7, 7′. If $A \cup B = U$ and $A \cap B = \varnothing$, then $B = \overline{A}$.
8, 8′. $\overline{\overline{A}} = A$.
9. $\overline{\varnothing} = U$.   9′. $\overline{U} = \varnothing$.
10. $A \cup A = A$.   10′. $A \cap A = A$.
11. $A \cup U = U$.   11′. $A \cap \varnothing = \varnothing$.
12. $A \cup (A \cap B) = A$.   12′. $A \cap (A \cup B) = A$.
13. $\overline{A \cup B} = \overline{A} \cap \overline{B}$.   13′. $\overline{A \cap B} = \overline{A} \cup \overline{B}$.

Some of the identities in Theorem 1.5.2 have well established names. For example, 10 and 10′ are the **idempotent laws,** 12 and 12′ are the **absorption laws,** and 13 and 13′ the **DeMorgan Laws.** The identities 7, 7′ and 8, 8′ are each numbered twice to emphasize that each is unchanged by the operation which converts it into its dual; such formulas are called **self-dual.** Note that 7, 7′ asserts that each set has a unique complement.

A remark about the form of the next theorem is in order. An assertion of the form, "The statements $R_1, R_2, \ldots, R_k$ are equivalent to one another," means "For all i and j, R_i iff R_j," which, in turn, is the case iff R_1 implies R_2 implies $R_3, \ldots, R_{k-1}$ implies R_k, and R_k implies R_1. The content of the theorem is that the inclusion relation for sets is definable in terms of union as well as in terms of intersection.

THEOREM 1.5.3. The following statements about sets A and B are equivalent to one another:

(I) $A \subseteq B$;
(II) $A \cap B = A$;
(III) $A \cup B = B$.

Proof. (I) implies (II). Assume that $A \subseteq B$. Since, for all A and B, $A \cap B \subseteq A$, it is sufficient to prove that $A \subseteq A \cap B$. But if $x \in A$, then $x \in B$ and, hence, $x \in A \cap B$. Hence $A \subseteq A \cap B$.

(II) implies (III). Assume $A \cap B = A$. Then

$A \cup B = (A \cap B) \cup B$
$\qquad\qquad = (A \cup B) \cap (B \cup B) = (A \cup B) \cap B = B.$

(III) implies (I). Assume that $A \cup B = B$. Then this and the identity $A \subseteq A \cup B$ imply $A \subseteq B$.

The principle of duality as formulated in terms of $\cup$, $\cap$, and $\bar{}$ does not directly apply to expressions in which $-$ and $\subseteq$ appear. One can cope with relative complements by using the unabbreviated form, namely $A \cap \bar{B}$, for $A - B$. For the inclusion relation we note that $A \subseteq B$ is equivalent to $A \cap B = A$ (Theorem 1.5.3), whose dual is $A \cup B = A$, which is equivalent to $A \supseteq B$. The foregoing observation is reversible and, hence, the notion of the dual of an expression and the principle of duality may be extended to include the case where the inclusion symbol is present by adding the clause that all inclusion symbols be reversed.

The characterization of $\cap$ and $\cup$ in terms of $\subseteq$ provided by Theorem 1.5.3 makes it possible to base the algebra of sets on the relation of inclusion. If this is done, the duality principle is then founded on simply the exchange of $\subseteq$ and $\supseteq$ in expressions. An advantage of this approach is that a principle of duality can be formulated for systems equipped with a relation for comparing elements and having some of the features enjoyed by the inclusion relation. (See §1.11 in this connection).

Examples

1.5.1. With the aid of the identities now available a great variety of complex expressions involving sets can be simplified, much as in elementary algebra. We give three illustrations.

(a) $\overline{A \cap \bar{B}} \cup B = \bar{A} \cup B \cup B = \bar{A} \cup B.$

(b) $(A \cap B \cap C) \cup (\bar{A} \cap B \cap C) \cup \bar{B} \cup \bar{C}$

$$= [(A \cup \bar{A}) \cap B \cap C] \cup \bar{B} \cup \bar{C}$$
$$= [U \cap B \cap C] \cup \overline{B \cap C}$$
$$= (B \cap C) \cup \overline{B \cap C}$$
$$= U.$$

(c) $(A \cap B \cap C \cap X) \cup (\bar{A} \cap C) \cup (\bar{B} \cap C) \cup (C \cap X)$

$$= (A \cap B \cap C \cap X) \cup [(\bar{A} \cup \bar{B} \cup X) \cap C]$$
$$= [(A \cap B \cap X) \cup \overline{A \cap B \cap X}] \cap C$$
$$= U \cap C$$
$$= C.$$

1.5.2. There is a theory of equations for the algebra of sets, and it differs considerably from that encountered in high-school algebra. As an illustration we shall discuss a method for solving a single equation in one "unknown." Such an equation may be described as one formed using $\cap$, $\cup$, $\bar{}$, and $=$ on symbols $A_1, A_2, \ldots, A_n$, and X, where the A's denote fixed subsets of some universal set U and X denotes a subset of U which is constrained only by the equation in which it appears. Using the algebra of sets, the problem is to determine under what conditions such an equation has a solution and then, assuming these are satisfied, to obtain all solutions. A recipe for this follows; the proof required in each step is left as an exercise (see Exercise 1.5.7).

Step I.   Two sets are equal iff their symmetric difference is equal to $\varnothing$. Hence, an equation in X is equivalent to one whose righthand side is $\varnothing$.

Step II.   An equation in X with righthand side $\varnothing$ is equivalent to one of the form

$$(A \cap X) \cup (B \cap \overline{X}) = \varnothing,$$

where A and B are free of X.

Step III.   The union of two sets is equal to $\varnothing$ iff each set is equal to $\varnothing$. Hence, the equation in Step II is equivalent to the pair of simultaneous equations

$$A \cap X = \varnothing, \; B \cap \overline{X} = \varnothing.$$

Step IV.   The above pair of equations, and hence the original equation, has a solution iff $B \subseteq \overline{A}$. In this event, any X, such that $B \subseteq X \subseteq \overline{A}$, is a solution.

We illustrate the foregoing by deriving necessary and sufficient conditions that the following equation have a solution:

$$X \cup C = D,$$
$$[(X \cup C) \cap \overline{D}] \cup [D \cap (\overline{X \cup C})] = \varnothing, \quad \text{(Step I)}$$
$$[(X \cup C) \cap \overline{D}] \cup [D \cap \overline{X} \cap \overline{C}] = \varnothing,$$
$$(X \cap \overline{D}) \cup (C \cap \overline{D}) \cup (D \cap \overline{X} \cap \overline{C}) = \varnothing,$$
$$(\overline{D} \cap X) \cup [(C \cap \overline{D}) \cap (X \cup \overline{X})] \cup (D \cap \overline{C} \cap \overline{X}) = \varnothing.$$

(The introduction of $X \cup \overline{X}$ in the preceding equation is discussed in Exercise 1.5.7.)

$$(\overline{D} \cap X) \cup (C \cap \overline{D} \cap X) \cup (C \cap \overline{D} \cap \overline{X}) \cup (D \cap \overline{C} \cap \overline{X}) = \varnothing,$$
$$\{[\overline{D} \cup (C \cap \overline{D})] \cap X\} \cup \{[(C \cap \overline{D}) \cup (D \cap \overline{C})] \cap \overline{X}\} = \varnothing,$$
$$(\overline{D} \cap X) \cup [(C + D) \cap \overline{X}] = \varnothing, \quad \text{(Step II)}$$
$$\overline{D} \cap X = \varnothing \text{ and } (C + D) \cap \overline{X} = \varnothing. \quad \text{(Step III)}$$

Thus, the original equation has a solution iff

$$C + D \subseteq D. \qquad\qquad \text{(Step IV)}$$

It is left as an exercise to show that this condition simplifies to $C \subseteq D$.

Exercises

1.5.1.   Prove that parts 3′, 4′, and 5′ of Theorem 1.5.1 are identities.

1.5.2.   Prove the unprimed parts of Theorem 1.5.2 using the membership relation. Try to prove the same results using only Theorem 1.5.1. In at least one such proof write out the dual of each step to demonstrate that a proof of the dual results.

1.5.3.   Using only the identities in Theorems 1.5.1 and 1.5.2, show that each of the following equations is an identity.

(a) $(A \cap B \cap X) \cup (A \cap B \cap C \cap X \cap Y) \cup (A \cap X \cap \overline{A})$
$= A \cap B \cap X.$

(b) $(A \cap B \cap C) \cup (\overline{A} \cap B \cap C) \cup \overline{B} \cup \overline{C} = U.$

(c) $(A \cap B \cap C \cap \overline{X}) \cup (\overline{A} \cap C) \cup (\overline{B} \cap C) \cup (C \cap X) = C.$

(d) $[(A \cap B) \cup (A \cap C) \cup (\overline{A} \cap \overline{X} \cap Y)] \cap [(A \cap \overline{B} \cap C) \cup (\overline{A} \cap \overline{X} \cap \overline{Y}) \cup (\overline{A} \cap B \cap Y)] = (A \cap B) \cup (\overline{A} \cap \overline{B} \cap \overline{X} \cap Y).$

1.5.4. Rework Exercise 1.4.9(a) using solely the algebra of sets developed in this section.

1.5.5. Let $A_1, A_2, \ldots, A_n$ be sets, and define S_k to be $A_1 \cup A_2 \cup \cdots \cup A_k$ for $k = 1, 2, \ldots, n$. Show that

$$\mathcal{C} = \{A_1, A_2 - S_1, A_3 - S_2, \ldots, A_n - S_{n-1}\}$$

is a disjoint collection of sets and that

$$S_n = A_1 \cup (A_2 - S_1) \cup \cdots \cup (A_n - S_{n-1}).$$

When is $\mathcal{C}$ a partition of S_n?

1.5.6. Prove that for arbitrary sets $A_1, A_2, \ldots, A_n (n \geq 2)$,

$$A_1 \cup A_2 \cup \cdots \cup A_n = (A_1 - A_2) \cup (A_2 - A_3) \cup \cdots \cup (A_{n-1} - A_n) \cup (A_n - A_1) \cup (A_1 \cap A_2 \cap \cdots \cap A_n).$$

1.5.7. Referring to Example 1.5.2, prove the following:

(a) For all sets A and B, $A = B$ iff $A + B = \varnothing$.

(b) An equation in X with righthand member $\varnothing$ can be reduced to one of the form $(A \cap X) \cup (B \cap \overline{X}) = \varnothing$. (Suggestion: Sketch a proof along these lines. First, apply the DeMorgan laws until only complements of individual sets appear. Then expand the resulting lefthand side by the distributive law 3, so as to transform it into the union of several terms T_i, each of which is an intersection of several individual sets. Next, if in any T_i neither X nor $\overline{X}$ appears, replace T_i by $T_i \cap (X \cup \overline{X})$ and expand. Finally, group together the terms containing X and those containing $\overline{X}$ and apply the distributive law 3'.)

(c) For all sets A and B, $A = B = \varnothing$ iff $A \cup B = \varnothing$.

(d) The equation $(A \cap X) \cup (B \cap \overline{X}) = \varnothing$ has a solution iff $B \subseteq \overline{A}$ and then any X such that $B \subseteq X \subseteq \overline{A}$ is a solution.

(e) An alternative form for solutions of the equation in part (d) is $X = (B \cup T) \cap \overline{A}$, where T is an arbitrary set.

1.5.8. Show that for arbitrary sets A, B, C, D, and X,

(a) $\overline{[(A \cap X) \cup (B \cap \overline{X})]} = (\overline{A} \cap X) \cup (\overline{B} \cap \overline{X})$;

(b) $[(A \cap X) \cup (B \cap \overline{X})] \cup [(C \cap X) \cup (D \cap \overline{X})]$
$= [(A \cup C) \cap X] \cup [(B \cup D) \cap \overline{X}]$;

(c) $[(A \cap X) \cup (B \cap \overline{X})] \cap [(C \cap X) \cup (D \cap \overline{X})]$
$= [(A \cap C) \cap X] \cup [(B \cap D) \cap \overline{X}].$

1.5.9. Using the results in Exercises 1.5.7 and 1.5.8, prove that the equation

$$(A \cap X) \cup (B \cap \overline{X}) = (C \cap X) \cup (D \cap \overline{X})$$

has a solution iff $B + D \subseteq \overline{A + C}$. In this event determine all solutions.

§1.6. Relations

In mathematics the word "relation" is used in the sense of relationship. The following partial sentences (or predicates) are examples of relations:

is less than,	is included in,
divides,	is a member of,
is congruent to,	is the mother of.

In this section the concept of a relation will be developed within the framework of set theory. The motivation for the forthcoming definition is this: A (binary) relation is used in connection with pairs of objects considered in a definite order. Further, a relation is concerned with the existence or nonexistence of some type of bond between certain ordered pairs. We infer that a relation provides a criterion for distinguishing some ordered pairs from others in the following sense. If a list of all ordered pairs for which the relation is pertinent is available, then with each may be associated "yes" or "no" to indicate that a pair is or is not in the given relation. Clearly, the same end is achieved by listing exactly all those pairs which are in the given relation. Such a list characterizes the relation. Thus the stage is set for defining a relation as a set of ordered pairs, and this is done as soon as the notion of an ordered pair is made precise.

Intuitively, an ordered pair is simply an entity consisting of two objects in a specified order. As the notion is used in mathematics, one relies on ordered pairs to have two properties: (i) given any two objects, x and y, there exists an object, which might be denoted by $\langle x, y \rangle$ and called the ordered pair of x and y, that is uniquely determined by x and y; (ii) if $\langle x, y \rangle$ and $\langle u, v \rangle$ are two ordered pairs, then $\langle x, y \rangle = \langle u, v \rangle$ iff $x = u$ and $y = v$. Now it is possible to define an object, indeed, a set, which has these properties: the **ordered pair** of x and y, symbolized by $\langle x, y \rangle$, is

$$\{\{x\}, \{x, y\}\},$$

that is, the two-element set one of whose members, $\{x, y\}$, is the unordered pair involved, and the other, $\{x\}$, determines which member of this unordered pair is to be considered as being "first." We shall now prove that, as defined, ordered pairs have the properties mentioned above.

THEOREM 1.6.1.   The ordered pair of x and y is uniquely determined by x and y. Moreover, if $\langle x,y \rangle = \langle u,v \rangle$, then $x = u$ and $y = v$.

Proof.   That x and y uniquely determine $\langle x,y \rangle$ follows from our assumption that a set is uniquely determined by its members. Turning to the more profound part of the proof, let us assume that $\langle x,y \rangle = \langle u,v \rangle$. We consider two cases:

(I) $u = v$. Then $\langle u,v \rangle = \{\{u\}, \{u,v\}\} = \{\{u\}\}$. Hence $\{\{x\}, \{x,y\}\}$ $= \{\{u\}\}$, which implies that $\{x\} = \{x,y\} = \{u\}$ and, in turn, that $x = u$ and $y = v$.

(II) $u \neq v$. Then $\{u\} \neq \{u,v\}$ and $\{x\} \neq \{u,v\}$. Since $\{x\} \in \{\{u\},$ $\{u,v\}\}$, it follows that $\{x\} = \{u\}$ and, hence, $x = u$. Since $\{u,v\} \in$ $\{\{x\}, \{x,y\}\}$ and $\{u,v\} \neq \{x\}$, we have $\{u,v\} = \{x,y\}$. Thus, $\{x\} \neq$ $\{x,y\}$, so, in turn, $x \neq y$ and $y \neq u$. Hence $y = v$.

We call x the **first coordinate** and y the **second coordinate** of the ordered pair $\langle x,y \rangle$. Ordered triples and, in general, ordered n-tuples may be defined in terms of ordered pairs. The **ordered triple** of x,y and z, symbolized by $\langle x,y,z \rangle$, is defined to be the ordered pair $\langle \langle x,y \rangle,z \rangle$. Assuming that ordered $(n-1)$-tuples have been defined, we take the **ordered n-tuple** of $x_1, x_2, \ldots, x_n$, symbolized by $\langle x_1,x_2, \ldots ,x_n \rangle$, to be $\langle \langle x_1,x_2, \ldots ,x_{n-1} \rangle,x_n \rangle$.

We return to our principal topic by defining a **binary relation** as a set of ordered pairs, that is, a set each of whose members is an ordered pair. If ρ is a relation, we write $\langle x,y \rangle \in \rho$ and $x\rho y$ interchangeably, and we say that x is ρ**-related to** y iff $x\rho y$. There are established symbols for various relations such as equality, membership, inclusion, congruence. Such familiar notation as $x = y$, $x < y$, and $x \equiv y$ is the origin of $x\rho y$ as a substitute for "$\langle x,y \rangle \in \rho$."

A natural generalization of a binary relation is that of an n**-ary relation** as a set of ordered n-tuples. The case $n = 2$ is, of course, the one for which we have agreed on the name "binary relation." Similarly, in place of 3-ary relation we shall say **ternary relation.**

Examples

1.6.1. $\{\langle 2,4 \rangle, \langle 7,3 \rangle, \langle 3,3 \rangle, \langle 2,1 \rangle\}$ as a set of ordered pairs is a binary relation. The fact that it appears to have no particular significance suggests that it is not worthwhile assigning a name to.

1.6.2. The relation "less than" for integers is $\{\langle x,y \rangle |$ for integers x and y, there is a positive integer z for which $x + z = y\}$. Symbolizing this relation in the traditional way, the statements "$2 < 5$" and "$\langle 2,5 \rangle \in <$" are synonymous (and true).

1.6.3. If μ symbolizes the relation of motherhood, then $\langle$ Jane, John $\rangle \in \mu$ means that Jane is the mother of John.

1.6.4. Human parenthood is an example of a ternary relation. If it is symbolized by ρ, then $\langle$ Elizabeth, Philip, Charles $\rangle \in \rho$ indicates that Elizabeth and Philip are the parents of Charles. Addition in $\mathbb{Z}$ is another ternary relation; writing "$5 = 2 + 3$" may be considered an alternative to asserting that $\langle 5,2,3 \rangle \in +$.

1.6.5. The cube-root relation for real numbers is $\{\langle x^{1/3}, x \rangle | x \in \mathbb{R}\}$. One member of this relation is $\langle 2,8 \rangle$.

1.6.6. In trigonometry the sine function is defined by way of a rule for associating with each real number a real number between -1 and 1. In practical applications one relies on a table in a handbook for values of this function for various arguments. Such a table is simply a compact way of displaying a set of ordered pairs. Thus, for practical purposes, the sine function is defined by the set of ordered pairs exhibited in a table (together with a rule concerning the extension of the table). We note that as such a table is designed to be read, it presents pairs of the form $\langle x, \sin x \rangle$; thereby the coordinates are interchanged from the order in which we have been writing them for relations in general. That is, for an arbitrary relation ρ we have interpreted $\langle a,b \rangle \in \rho$ as meaning that a is ρ-related to b, whereas the presence of $\langle \pi/2, 1 \rangle$ in a table for the sine function is intended to convey the information that the second coordinate is sine-related to (is the sine of) the first coordinate.

Later we shall find extensive applications for ternary relations, but our present interest is in binary relations, which we shall abbreviate to simply "relations" if no confusion can result. If ρ is a relation, then the **domain** of ρ, symbolized by D_ρ, is

$$\{x| \text{ for some } y, \langle x,y \rangle \in \rho\},$$

the **range** of ρ, symbolized by R_ρ, is

$$\{y| \text{ for some } x, \langle x,y \rangle \in \rho\},$$

and the **field** of ρ, symbolized F_ρ, is the union of its domain and range. That is, the domain of ρ is the set whose members are the first coordinates of members of ρ, and the range of ρ is the set whose members are the second coordinates of members of ρ. For example, the domain and range of the inclusion relation for subsets of a set U are each equal to $\mathcal{P}(U)$. Again, for human beings, the domain of the relation of motherhood is the set of all mothers, and the range is the set of all people.

One of the simplest types of relations is the set of all pairs $\langle x, y \rangle$, such that x is a member of some fixed set X and y is a member of some fixed set Y. This relation is the **cartesian product**, $X \times Y$, of X and Y. Thus,

$$X \times Y = \{\langle x, y \rangle | x \in X \text{ and } y \in Y\}.$$

It is evident that a relation ρ is a subset of any cartesian product $X \times Y$, such that $X \supseteq D_\rho$ and $Y \supseteq R_\rho$. If ρ is a relation and $\rho \subseteq X \times Y$, then ρ is referred to as a **relation from** X **to** Y. If ρ is a relation from X to Y and $Z \supseteq X \cup Y$, then ρ is a relation from Z to Z. A relation from Z to Z will be called a **relation in** Z. Such terminologies as "a relation from X to Y" and "a relation in Z" stem from the possible application of a relation to distinguish certain ordered pairs of objects from others. If X is a set, then $X \times X$ is a relation in X which we shall call the **universal relation in** X; this is a suggestive name, since, for each pair x, y of elements in X, we have $x(X \times X)y$. At the other extreme is the **void relation** in X, consisting of the empty set. Intermediate is the **identity relation** in X, symbolized by ι or ι_X, which is $\{\langle x, x \rangle | x \in X\}$; for x, y in X, clearly, $x\iota_X y$ iff $x = y$.

If ρ is a relation and A is a set, then $\rho[A]$ is defined to be

$$\{y| \text{ for some } x \text{ in } A, x\rho y\}.$$

This set is suggestively called the set of ρ-**relatives** of elements of A. Clearly, $\rho[D_\rho] = R_\rho$, and, if A is any set, $\rho[A] \subseteq R_\rho$.

Examples

1.6.7. If $Y \neq \varnothing$, then $D_{X \times Y} = X$, and if $X \neq \varnothing$, then $R_{X \times Y} = Y$.

1.6.8. The basis for plane analytic geometry is the assumption that the points of the Euclidean plane can be paired with the members of $\mathbb{R} \times \mathbb{R}$, the set of ordered pairs of real numbers. Thereby the study of plane geometric configurations may be replaced by that of subsets of $\mathbb{R} \times \mathbb{R}$, that is, relations in $\mathbb{R}$. For geometric configurations which are likely to be of interest, one can anticipate that the defining property of the associated relation in $\mathbb{R}$ will be an algebraic equation in x and y, or an inequality involving x and y, or some combination of equations and inequalities. In this event it is standard practice to take the defining property of the relation associated with a configuration as a description of the configuration and omit any explicit mention of the relation. For example, "the line with equation $y = 2x + 1$" is shorthand for "the set of points which are associated with $\{\langle x, y \rangle \in \mathbb{R} \times \mathbb{R} | y = 2x + 1\}$." Again, "the region defined by $y < x$" is intended to refer to the set of points associated with $\{\langle x, y \rangle \in \mathbb{R} \times \mathbb{R} | y < x\}$. As a further example,

$$x \leq 0 \text{ and } y \geq 0 \text{ and } y \leq 2x + 1$$

serves as a definition of a triangle-shaped region in the plane, as the reader can verify.

If relations in $\mathbb{R}$, instead of sets of points in the plane, are the primary objects of study, then the set of points corresponding to the members of a relation is called the **graph** of the relation (or of the defining property of the relation). Figures 1.9 to 1.12 show the graphs for four relations. When the graph includes a region of the plane, this is indicated by shading.

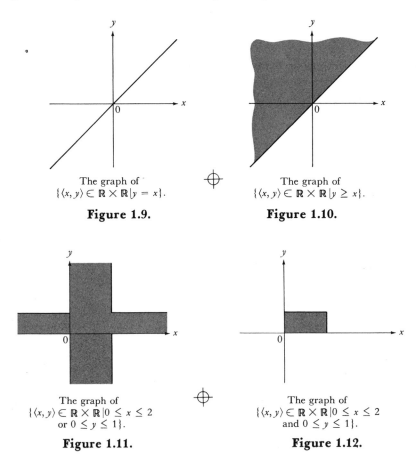

The graph of
$\{\langle x, y \rangle \in \mathbb{R} \times \mathbb{R} \,|\, y = x\}$.

Figure 1.9.

The graph of
$\{\langle x, y \rangle \in \mathbb{R} \times \mathbb{R} \,|\, y \geq x\}$.

Figure 1.10.

The graph of
$\{\langle x, y \rangle \in \mathbb{R} \times \mathbb{R} \,|\, 0 \leq x \leq 2$
or $0 \leq y \leq 1\}$.

Figure 1.11.

The graph of
$\{\langle x, y \rangle \in \mathbb{R} \times \mathbb{R} \,|\, 0 \leq x \leq 2$
and $0 \leq y \leq 1\}$.

Figure 1.12.

If ρ is the relation in $\mathbb{R}$ with $0 \leq x \leq 2$ as defining property and σ is the relation in $\mathbb{R}$ with $0 \leq y \leq 1$ as defining property, then the relation graphed in Figure 1.11 is equal to $\rho \cup \sigma$, and the relation graphed in Figure 1.12 is $\rho \cap \sigma$. Thus, Figures 1.11 and 1.12 illustrate the remarks that the graph of the union of two relations, ρ and σ, is the union of the graph of ρ and the graph of σ, and the graph of $\rho \cap \sigma$ is the intersection of the graphs of ρ and σ.

1.6.9. Let ρ be the relation "is the father of." If A is the set of all men now living in the United States, then $\rho[A]$ is the set of all people whose fathers now live in the United States. If $B = \{$Adam, Eve$\}$, then $\rho[B] = \{$Cain, Abel$\}$.

The earlier algebra of sets is, of course, applicable to relations. However, the operations of union, intersection, and complementation are of no particular significance for the theory of relations. Two operations which are useful are defined next. The **relative product** of the relations ρ and σ, symbolized $\rho|\sigma$, is (the relation)

$$\{\langle x,y\rangle| \text{ for some } z, \langle x,z\rangle \in \rho \text{ and } \langle z,y\rangle \in \sigma\}.$$

If ρ is a relation, the **inverse** or **converse** of ρ, symbolized ρ^{-1}, is (the relation)

$$\{\langle x,y\rangle|\langle y,x\rangle \in \rho\}.$$

Applications of these operations will be given as we study the various classes of relations in the remainder of this chapter.

Exercises

1.6.1. Show that if $\langle x,y,z\rangle = \langle u,v,w\rangle$, then $x = u$, $y = v$, and $z = w$.

1.6.2. Write the members of $\{1,2\} \times \{2,3,4\}$. What are the domain and range of this relation? What is its graph?

1.6.3. State the domain and the range of each of the following relations, and then draw its graph.

    (a) $\{\langle x,y\rangle \in \mathbb{R} \times \mathbb{R}|x^2 + 4y^2 = 1\}$.
    (b) $\{\langle x,y\rangle \in \mathbb{R} \times \mathbb{R}|x^2 = y^2\}$.
    (c) $\{\langle x,y\rangle \in \mathbb{R} \times \mathbb{R}||x| + 2|y| = 1\}$.
    (d) $\{\langle x,y\rangle \in \mathbb{R} \times \mathbb{R}|x^2 + y^2 < 1 \text{ and } x > 0\}$.
    (e) $\{\langle x,y\rangle \in \mathbb{R} \times \mathbb{R}|y \geq 0 \text{ and } y \leq x \text{ and } x + y \leq 1\}$.

1.6.4. Write the relation in Exercise 1.6.3(c) as the union of four relations and that in Exercise 1.6.3(e) as the intersection of three relations.

1.6.5. The formation of the cartesian product of two sets is a binary operation for sets. Show by examples that this operation is neither commutative nor associative.

1.6.6. Let β be the relation "is a brother of," and let σ be the relation "is a sister of." Describe $\beta \cup \sigma$, $\beta \cap \sigma$, and $\beta - \sigma$.

1.6.7. Let β and σ have the same meaning as in Exercise 1.6.6. Let A be the set of students now in the reader's school. What is $\beta[A]$? What is $(\beta \cup \sigma)[A]$?

1.6.8. Prove that if A, B, C, and D are sets, then $(A \cap B) \times (C \cap D) = (A \times C) \cap (B \times D)$. Deduce that the cartesian multiplication of sets distributes over the operation of intersection, that is, that $(A \cap B) \times C = (A \times C) \cap (B \times C)$ and $A \times (B \cap C) = (A \times B) \cap (A \times C)$ for all A, B, and C.

1.6.9. Exhibit four sets A, B, C, and D for which $(A \cup B) \times (C \cup D) \neq (A \times C) \cup (B \times D)$.

1.6.10. In spite of the result in the preceding exercise, cartesian multiplication distributes over the operation of union. Prove this.

1.6.11. Investigate whether union and intersection distribute over cartesian multiplication.

1.6.12. Prove that if A, B, and C are sets such that $A \neq \varnothing$, $B \neq \varnothing$, and $(A \times B) \cup (B \times A) = C \times C$, then $A = B = C$.

1.6.13. Derive the following identities for relations ρ, σ, and τ.

(a) $\rho|(\sigma|\tau) = (\rho|\sigma)|\tau$.

(b) $\rho|(\sigma \cup \tau) = (\rho|\sigma) \cup (\rho|\tau)$ and $(\rho \cup \sigma)|\tau = (\rho|\tau) \cup (\sigma|\tau)$.

(c) $\rho|(\sigma \cap \tau) \subseteq (\rho|\sigma) \cap (\rho|\tau)$ and $(\rho \cap \sigma)|\tau \subseteq (\rho|\tau) \cap (\sigma|\tau)$.

(d) $(\rho|\sigma) \cap \tau = \varnothing$ iff $(\rho^{-1}|\tau) \cap \sigma = \varnothing$.

1.6.14. Show that the inclusions in Exercise 1.6.13(c) cannot be replaced by equalities.

1.6.15. Derive the following identities for relations ρ and σ.

(a) $(\rho \cup \sigma)^{-1} = \rho^{-1} \cup \sigma^{-1}$.

(b) $(\rho^{-1})^{-1} = \rho$.

(c) $(\rho|\sigma)^{-1} = \sigma^{-1}|\rho^{-1}$.

§1.7. Equivalence Relations

A relation ρ may be considered as given by itself or as a relation in a set X. If ρ is a relation in X, it may be that ρ was given first and X was obtained in some way from the domain and the range of ρ; or else X was given and ρ was introduced for the purpose of imposing some structure on X or was obtained in some way from the known structure of X. In a study of the theory of relations, it may seem more natural to consider relations as such, rather than relations in a set. In applications, on the other hand, relations often occur as relations in a set. The various types of relations which are introduced in the remainder of this chapter are considered from both points of view.

We begin by naming several properties, some or all of which are possessed by such familiar relations as equality, divisibility, and inclusion. A relation ρ is **reflexive on** X (a set) iff $x\rho x$ for each x in X; ρ is **reflexive** iff it is reflexive on $D_\rho \cup R_\rho$, the so-called **field**, F_ρ, of ρ. A relation ρ is **symmetric** iff $x\rho y$ implies $y\rho x$, and it is **transitive** iff $x\rho y$ and $y\rho z$ imply $x\rho z$. Relations having these three properties occur so frequently in mathematics that they have acquired a name. A relation ρ is an **equivalence relation** (on X) iff it is reflexive (on X), symmetric, and transitive. Observe that the foregoing sentence gives two definitions—that of an equivalence relation and that of an equivalence relation on a set X. The assumption of reflexivity for an equivalence relation ρ is redundant, since from the symmetry and transitivity of ρ follows its reflexivity. However, a relation ρ may be symmetric and transitive yet not be an equivalence relation on some set X; see Example 1.7.8 below. Further, we note that if ρ is an equivalence relation, then $D_\rho = R_\rho = F_\rho$, and if ρ is an equivalence relation on X, then $D_\rho = R_\rho = X$. Formulations of the above definitions in terms of relative products and inverses are given in the exercises.

Examples

Each of the relations defined in 1.7.1-1.7.7 is an equivalence relation on the accompanying set.

1.7.1. Equality in a collection of sets.

1.7.2. The geometric notion of similarity in the set of all triangles of the Euclidean plane.

1.7.3. The relation of congruence modulo n in $\mathbb{Z}$. This relation is defined for a nonzero integer n as follows: x is congruent to y, symbolized $x \equiv y \pmod{n}$, iff n divides $x - y$.

1.7.4. The relation $\sim$ in the set of all ordered pairs of positive integers where $\langle x, y \rangle \sim \langle u, v \rangle$ iff $xv = yu$.

1.7.5. The relation of parallelism in the set of lines in the Euclidean plane.

1.7.6. The relation of having the same number of members in a collection of finite sets.

1.7.7. The relation of "living in the same house" in the set of people of the United States.

1.7.8. The relation $\rho = \{\langle 1,1 \rangle, \langle 1,2 \rangle, \langle 2,1 \rangle, \langle 2,2 \rangle\}$ is an equivalence relation. It is *not* an equivalence relation on $\{1,2,3\}$.

Example 1.7.7 illustrates, in familiar terms, the central feature of any equivalence relation on a set: it divides the population into pairwise disjoint subsets, in this case the sets of people who live in the same house. Let us establish our contention in general. If ρ is an equivalence relation on the set X, then a subset A of X is an **equivalence class** (ρ-equivalence class) iff there is a member x of A such that A is equal to the set of all y for which $x\rho y$. Thus, A is an equivalence class iff there exists an x in X such that $A = \rho[\{x\}]$, the set of ρ-relative of x. If there is no ambiguity about the relation at hand, the set of all ρ-relatives of x in X will be abbreviated $[x]$ and called the equivalence class generated by x. Two basic properties of equivalence classes are the following:

(I) $x \in [x]$;

(II) if $x\rho y$, then $[x] = [y]$.

The first is a consequence of the reflexivity of an equivalence relation. To prove the second, assume that $x \rho y$. Then $[y] \subseteq [x]$ since $z \in [y]$ (which means that $y \rho z$) together with $x \rho y$ and the transitivity of ρ yield $x \rho z$ or $z \in [x]$. The symmetry of ρ may be used to conclude the reverse inclusion, and the equality of $[x]$ and $[y]$ follows.

Now property (I) implies that each member of X is a member of an equivalence class, and (II) implies that two equivalence classes are either disjoint or equal since if $z \in [x]$ and $z \in [y]$, then $[x] = [z]$, $[y] = [z]$, and hence $[x] = [y]$. Recalling the definition of a partition of a nonempty set, we conclude that the collection of distinct ρ-equivalence classes is a partition of X. This proves the first assertion in the following theorem.

THEOREM 1.7.1. Let ρ be an equivalence relation on X. Then the collection of distinct ρ-equivalence classes is a partition of X. Conversely, if $\mathcal{P}$ is a partition of X, and a relation ρ is defined by $a \rho b$ iff there exists A in $\mathcal{P}$ such that $a, b \in A$, then ρ is an equivalence relation on X. Moreover, if an equivalence relation ρ determines the partition $\mathcal{P}$ of X, then the equivalence relation defined by $\mathcal{P}$ is equal to ρ. Conversely, if a partition $\mathcal{P}$ of X determines the equivalence relation ρ, then the partition of X defined by ρ is equal to $\mathcal{P}$.

Proof. To prove the second statement, let $\mathcal{P}$ be a partition of X. The relation ρ which is proposed is symmetric from its definition. If $a \in X$, there exists A in $\mathcal{P}$ with $a \in A$, so that ρ is reflexive. To show the transitivity of ρ, assume that $a \rho b$ and $b \rho c$. Then there exists A in $\mathcal{P}$ with $a, b \in A$, and there exists B in $\mathcal{P}$ with $b, c \in B$. Since $b \in A$ and $b \in B$, $A = B$. Hence $a \rho c$.

To prove the next assertion, assume that an equivalence relation ρ on X is given, that it determines the partition $\mathcal{P}$ of X and, finally, that $\mathcal{P}$ determines the equivalence relation ρ^*. We show that $\rho = \rho^*$. Assume that $\langle x, y \rangle \in \rho$. Then $x, y \in [x]$ and $[x] \in \mathcal{P}$. By virtue of the definition of ρ^* it follows that $x \rho^* y$ or $\langle x, y \rangle \in \rho^*$. Conversely, given $\langle x, y \rangle \in \rho^*$, there exists A in $\mathcal{P}$ with $x, y \in A$. But A is a ρ-equivalence class, and hence $x \rho y$ or $\langle x, y \rangle \in \rho$. Thus, $\rho = \rho^*$.

The last part of the theorem is left as an exercise.

To illustrate part of the above theorem let us examine the equivalence relation of congruence modulo n on $\mathbb{Z}$ which was defined in Example 1.7.3. Now we assume that n is a positive integer. An equivalence class consists of all numbers $a + kn$ with k in $\mathbb{Z}$. Clearly, therefore, $[0]$,

[1], . . . , [n − 1] are distinct classes. There are no others, since any integer a can be written in the form $a = qn + r$, $0 \le r < n$, and hence $a \in [r]$. A class of congruent numbers is often called a **residue class** modulo n. The collection of residue classes modulo n will be denoted by $\mathbb{Z}_n$. We can use this example to emphasize the fact that, for any equivalence relation ρ, an equivalence class is defined by any one of its members, since if $x\rho y$, then $[x] = [y]$. Thus, $[0] = [n] = [2n]$, and so on, and $[1] = [n + 1] = [1 − n]$, and so on.

If ρ is an equivalence relation on X, we shall denote the partition of X induced by ρ by X/ρ (read "X modulo ρ") and call it the **quotient set** of X by ρ. The significance of the partition of a set X accompanying an arbitrary equivalence relation ρ on X is best realized by comparing ρ with the extreme equivalence relation on X of identity. We classify identity on X as an extreme equivalence relation because the only element equal to a given element is itself. That is, the partition of X determined by identity is the finest possible—the equivalence class generated by x consists of x alone. In contrast, for two elements to be ρ-equivalent they must merely have a single likeness in common, namely, that characterized by ρ. A ρ-equivalence class consists of all elements of X which are indiscernible with respect to ρ. That is, an arbitrary equivalence relation on X defines a generalized form of equality on X. On turning from the elements of X to the ρ-equivalence classes we have the effect of identifying any two elements which are ρ-equivalent. If ρ happens to preserve various structural features of X (assuming it has such), these may appear in simplified form in X/ρ because of the identification of elements which accompanies the transition to X/ρ. Examples of this arise quite naturally later.

Among the applications of equivalence relations in mathematics is that of formalizing mathematical notions or, as one often says, formulating definitions by abstraction. The essence of this technique is defining a notion as the set of all objects which one intends to have qualify for the notion. This seems incestuous on the surface, but in practice it serves very nicely. For example, let us consider the problem of defining the rational numbers in terms of the integers. We begin by considering the set F of all fractions, that is, the set of all ordered pairs $\langle p,q \rangle$ where p and q are integers and $q \ne 0$. Different fractions, for example $\langle 2,3 \rangle$ and $\langle 4,6 \rangle$, may express the same rational number. In arithmetic one learns the criterion: $\langle p,q \rangle$ and $\langle r,s \rangle$ express the same rational number if $ps = qr$. This leads us to define a relation, $\sim$, in F by

$$\langle p,q \rangle \sim \langle r,s \rangle \quad \text{iff} \quad ps = qr.$$

This is a criterion for indiscernibility in the set F. Since, as is easily checked, $\sim$ is an equivalence relation on F, it induces a partition of F, and in an equivalence class we have the abstraction of the property common to all its members. The familiar symbol p/q emerges as an abbreviation for the equivalence class $[\langle p,q \rangle]$. That an equivalence class is defined by each of its members implies that any other symbol r/s, where $\langle r,s \rangle \in [\langle p,q \rangle]$, may be taken as a name for the same rational number. For example, the statement $2/3 = 4/6$ is true because $2/3$ and $4/6$ are names of the same rational number. Stated another way, in the transition from F (the fractions) to $F/\sim$ (the rational numbers), the equivalence relation $\sim$ on F gives rise to the equality relation $=$ of $F/\sim$: two rationals x and y are the same ($x = y$) iff for any $\langle p,q \rangle$ in x and $\langle r,s \rangle$ in y, $\langle p,q \rangle \sim \langle r,s \rangle$.

Another instance of definition by abstraction is that of direction based on the equivalence relation of parallelism: a direction is an equivalence class of parallel rays. The notion of shape may be conceived in a like fashion: geometric similarity is an equivalence relation on the set of figures in the Euclidean plane, and a shape may be defined as an equivalence class under similarity.

So far, the fundamental result concerning an equivalence relation ρ—that the collection of all distinct ρ-equivalence classes is disjoint and $x\rho y$ iff x and y are members of the same equivalence class—has been employed solely in connection with applications of equivalence relations. It can also be made the basis of a characterization of equivalence relations among relations in general. This is done next.

THEOREM 1.7.2. A relation ρ is an equivalence relation iff there exists a disjoint collection $\mathcal{P}$ of nonempty sets such that

$$\rho = \{\langle x,y \rangle | \text{ for some } C \text{ in } \mathcal{P}, \langle x,y \rangle \in C \times C\}.$$

Proof. Assume that ρ is an equivalence relation. Then it is an equivalence relation on its field F_ρ. The collection of distinct ρ-equivalence classes, as a partition of F_ρ, is disjoint, and we contend that with this choice for $\mathcal{P}$, ρ has the structure described in the theorem. We show first that $\{\langle x,y \rangle | \text{ for some } C \text{ in } \mathcal{P}, \langle x,y \rangle \in C \times C\} \subseteq \rho$. Assume that $\langle x,y \rangle$ is a member of the set on the left side of the inclusion sign. Then there exists an equivalence class $[z]$ with $x, y \in [z]$. Then $z\rho x$ and $z\rho y$, and hence $x\rho y$, which means that $\langle x,y \rangle \in \rho$. To show the reverse inclusion, assume that $\langle x,y \rangle \in \rho$. Then $x, y \in [x]$, and hence $\langle x,y \rangle \in [x] \times [x]$.

The proof of the converse is straightforward and is left as an exercise.

Exercises

1.7.1. If ρ is a relation in $\mathbb{R}^+$, then its graph is a set of points in the first quadrant of a coordinate plane. What is the characteristic feature of such a graph if: (a) ρ is reflexive on $\mathbb{R}^+$, (b) ρ is symmetric, (c) ρ is transitive?

1.7.2. Using the results of Exercise 1.7.1, try to formulate a compact characterization of the graph of an equivalence relation on $\mathbb{R}^+$.

1.7.3. The collection of sets $\{\{1,3,4\}, \{2,7\}, \{5,6\}\}$ is a partition of $\{1,2,3,4,5,6,7\}$. Draw the graph of the accompanying equivalence relation.

1.7.4. Let ρ and σ be equivalence relations. Prove that $\rho \cap \sigma$ is an equivalence relation.

1.7.5. Let ρ be an equivalence relation on X and let Y be a set. Show that $\rho \cap (Y \times Y)$ is an equivalence relation on $X \cap Y$.

1.7.6. Give an example of:
   (a) a relation which is reflexive and symmetric but not transitive;
   (b) a relation which is reflexive and transitive but not symmetric;
   (c) a relation which is symmetric and transitive but not reflexive on some set.

1.7.7. Complete the proof of Theorem 1.7.1.

1.7.8. Each equivalence relation on a set X defines a partition of X according to Theorem 1.7.1. What equivalence yields the finest partition? The coarsest partition?

1.7.9. Complete the proof of Theorem 1.7.2.

1.7.10. Let ρ be a relation which is transitive and reflexive on the set A. For $a, b \in A$, define $a \sim b$ iff $a\rho b$ and $b\rho a$.
   (a) Show that $\sim$ is an equivalence relation on A.
   (b) For $[a], [b] \in A/\sim$, define $[a]\rho'[b]$ iff $a\rho b$. Show that this definition is independent of a and b in the sense that if $a' \in [a]$, $b' \in [b]$, and $a\rho b$, then $a'\rho b'$.
   (c) Show that ρ' is reflexive and transitive. Further, show that if $[a]\rho'[b]$ and $[b]\rho'[a]$, then $[a] = [b]$.

1.7.11. In the set $\mathbb{Z}^+ \times \mathbb{Z}^+$ define $\langle a,b\rangle \sim \langle c,d\rangle$ iff $a + d = b + c$. Show that $\sim$ is an equivalence relation on this set. Indicate the graph of $\mathbb{Z}^+ \times \mathbb{Z}^+$, and describe the $\sim$-equivalence classes.

1.7.12. Let ρ be a relation. Show that:
   (a) ρ is reflexive on a set X iff $\rho \supseteq i_X$;
   (b) ρ is symmetric iff $\rho^{-1} = \rho$;
   (c) ρ is transitive iff $\rho|\rho \subseteq \rho$.

1.7.13. Show that a relation ρ is an equivalence relation iff $\rho|\rho^{-1} = \rho$.

§1.8. Functions

It is possible to define the concept of function in terms of notions already introduced. Such a definition is based on the common part of the discussions about functions to be found in many elementary texts, namely, the definition of the graph of a function as a set of ordered pairs. Once it is recognized that there is no information about a function which cannot be derived from its graph, there is no need to distinguish between a function and its graph. As such, it is reasonable to base a definition on just that feature of a set of ordered pairs which would qualify it to be a graph of a function. This we do by agreeing that a **function** is a relation such that no two distinct members have the same first coordinate. Thus, f is a function iff it meets the following requirements:

(I) The members of f are ordered pairs.

(II) If $\langle x,y \rangle$ and $\langle x,z \rangle$ are members of f, then $y = z$.

Examples

1.8.1. $\{\langle 1,2 \rangle, \langle 2,2 \rangle, \langle \text{Roosevelt, Churchill} \rangle\}$ is a function with domain $\{1, 2, \text{Roosevelt}\}$ and range $\{2, \text{Churchill}\}$.

1.8.2. The relation $\{\langle 1,2 \rangle, \langle 1,3 \rangle, \langle 2,2 \rangle\}$ is not a function, since the distinct members $\langle 1,2 \rangle$ and $\langle 1,3 \rangle$ have the same first coordinate.

1.8.3. The relation $\{\langle x, x^2 + x + 1 \rangle | x \in \mathbb{R}\}$ is a function, because if $x = u$, then $x^2 + x + 1 = u^2 + u + 1$.

1.8.4. The relation $\{\langle x^2, x \rangle | x \in \mathbb{R}\}$ is not a function, because both $\langle 1,1 \rangle$ and $\langle 1, -1 \rangle$ are members.

Synonyms for the word "function" are numerous and include **transformation, map** or **mapping, correspondence,** and **operator.*** If f is a function and $\langle x,y \rangle \in f$, so that xfy, then x is an **argument** of f. There is a great variety of terminology for y; for example, the **value** of f at x, the **image** of x under f, the element into which f **carries** x. There are also various symbols for y: xf, $f(x)$ (or, more simply, fx), x^f. The notation $f(x)$ may be regarded as an abbreviation for the sole member of $f[\{x\}]$, the set of f-relatives of x. In these terms the characteristic feature of a function among relations in general is that each member of the domain of a function has a single relative.

The student must accustom himself to these various notations, since he will find that all are used. In this book definitions and theorems pertaining to functions will consistently be phrased using the notation $f(x)$, or fx, for the (unique) correspondent of x in a function f. The notation $f[A]$ for $\{y|$ for some x in A, $\langle x,y \rangle \in f\}$ is in harmony with this. However, in applications of functions we shall use a variety of notations. When it

---

* In certain contexts one of these names may be selected for functions of a class that possess one or more special properties.

is more convenient to use xf in place of $f(x)$, then $[A]f$ will be used in place of $f[A]$. If x^f is used in place of $f(x)$, then $[A]^f$ or A^f will be used in place of $f[A]$.

Since functions are sets, the definition of equality of functions is at hand: Two functions f and g are equal iff they have the same members. It is clear that this may be rephrased $f = g$ iff $D_f = D_g$ and $f(x) = g(x)$ for each x in the common domain. Consequently, a function may be defined by specifying its domain and the value of the function at each member of its domain. The second part of this type of definition is, then, in the nature of a rule. For example, an alternative definition of the function $\{\langle x, x^2 + x + 1 \rangle | x \in \mathbb{R}\}$ is the function f with $\mathbb{R}$ as domain and such that $f(x) = x^2 + x + 1$. Another notation for functions is associated with the American logician Alonzo Church (b. 1903), although it dates back to Frege. It is commonly called (Church's) **lambda notation.** If we abbreviate by "$f(x)$" an expression containing "x," which indicates the value of a function when the argument has value x, we write

$$\lambda x[f(x)]$$

to designate the function itself. Thus, $\lambda x[x^2 + x + 1]$ designates the function discussed above. If a is a member of the domain of the function, then its value at a is written $\lambda x[f(x)]a$, so that

$$\lambda x[f(x)]a = f(a).$$

A variant on the lambda notation to designate a function is to use a barred arrow ($\mapsto$) to relate an argument of the function and the associated value. In this notation, the function f defined above is designated by

$$x \mapsto x^2 + x + 1 \qquad \text{for } x \in \mathbb{R}.$$

When a function is defined by specifying its domain and its value at each member of the domain, the range of the function may not be evident. For the function f defined above, a computation is required to conclude that $R_f = \{x \in \mathbb{R} | x \geq 3/4\}$. On the other hand it is obvious that $R_f \subseteq \mathbb{R}$. In general, one can anticipate difficulty in determining the range of a function, but no difficulty in determining some set that includes the range. Thus it is convenient to have available the following terminology. If f is a function with domain X, and Y is a set that includes the range of f, then we shall say that f is a function **on** (or **from**) X **into** (or **to**) Y, or that f **maps** X **into** Y, or that f is **defined on** X **with values in** Y. This circumstance is often expressed in symbols by

$$f\colon X \longrightarrow Y \text{ or } X \stackrel{f}{\longrightarrow} Y.$$

If $Y = R_f$, then f is said to be **surjective** or **onto** Y, or a **surjection.** We note that the property of being onto (or surjective) is not a property of f alone but one of f and Y.

The set of all functions on X into Y is a subset of $\mathcal{P}(X \times Y)$, which we shall symbolize by Y^X. If X is empty, then Y^X consists of only one member, namely, the empty subset of $X \times Y$. This is the only subset of $X \times Y$, since when X is empty so is $X \times Y$. If Y is empty and X is non-empty, then Y^X is empty.

If $f\colon X \longrightarrow Y$, and if $A \subseteq X$, then $f \cap (A \times Y)$ is a function on A into Y (called the **restriction** of f to A and abbreviated $f{\restriction}A$). Explicitly, $f{\restriction}A$ is the function on A such that $(f{\restriction}A)(a) = f(a)$ for a in A. A function g is the restriction of a function f to some subset of the domain of f iff the domain of g is a subset of the domain of f and $g(x) = f(x)$ for $x \in D_g$; in other words, $g \subseteq f$. Complementary to the definition of a restriction, the function f is an **extension** of a function g iff $g \subseteq f$. In order to present an example of a restriction of a function, we recall the earlier definition of the identity relation ι_X in X. Clearly, this relation is a function, and hence, in keeping with our current designation of function by lowercase English letters, we shall designate it by i or i_X. We shall call i_X the **identity map** on X. If $A \subseteq X$, then $i_X|A = i_A$; the function $i_A\colon A \longrightarrow X$ is called the **inclusion map** of A into X.

A function is called **injective** or **one-to-one** or an **injection** iff it maps distinct elements onto distinct elements. That is, a function f is injective iff

$$x_1 \neq x_2 \text{ implies } f(x_1) \neq f(x_2)$$

or, what is logically equivalent,

$$f(x_1) = f(x_2) \text{ implies } x_1 = x_2.$$

As an example, the function on $\mathbb{R}$ given by $x \mapsto 2x + 1$ is injective since $2x_1 + 1 = 2x_2 + 1$ implies $x_1 = x_2$.

A function is called **bijective** or a **bijection** iff it is both injective and surjective. Since surjectivity is not, strictly speaking, a property of a function, neither is bijectivity. If $f\colon X \longrightarrow Y$ is bijective, then it effects a pairing of the elements of X with those of Y upon matching $f(x)$ in Y with x in X. Indeed, since f is a function, $f(x)$ is a uniquely determined element of Y; since f is surjective, each y in Y is matched with some x, and since f is injective, each y in Y is matched with only one x. Because of the symmetrical situation that a bijection $f\colon X \longrightarrow Y$ portrays, it is often called a **one-to-one correspondence between** X **and** Y. Also, two sets so related by some function are said to be in one-to-one correspondence.

Examples

1.8.5. The familiar exponential function is a function on $\mathbb{R}$ into $\mathbb{R}$, symbolized

$$f\colon \mathbb{R} \longrightarrow \mathbb{R} \text{ with } f(x) = e^x.$$

We can also say, more precisely, that f is a function on $\mathbb{R}$ onto $\mathbb{R}^+$. This function is a bijection, so that $\mathbb{R}$ and $\mathbb{R}^+$ are in one-to-one correspondence. In general, if $f\colon X \longrightarrow Y$, then f is a function on X onto $f[X]$, that is, onto the range of f.

1.8.6. $\{a,b,c\}^{\{1,2\}}$ is the set of all functions on $\{1,2\}$ into $\{a,b,c\}$. One member of this set is $\{\langle 1,a\rangle, \langle 2,c\rangle\}$.

1.8.7. If A and B are sets having the same number of elements, they may be placed in one-to-one correspondence. Then it is an easy matter to show that for any set X, A^X and B^X may be placed in one-to-one correspondence. This being the case, it is customary to denote the set of all functions on X into any set of n elements by n^X. Thus, 2^X denotes the set of all functions on X into a set of two elements, which we will ordinarily take to be $\{0,1\}$. If $A \subseteq X$, then one member of 2^X is the function χ_A defined as follows:

$$\chi_A(x) = 1 \text{ if } x \in A, \text{ and } \chi_A(x) = 0 \text{ if } x \in X - A.$$

We call χ_A the **characteristic function** of A. Now let us define a function f on $\mathcal{P}(X)$ into 2^X by taking as the image of a subset A of X [that is, a member of $\mathcal{P}(X)$] the characteristic function of A (which is a member of 2^X). It is left as an exercise to prove that f is a one-to-one correspondence between $\mathcal{P}(X)$ and 2^X. It is customary to regard $\mathcal{P}(X)$ and 2^X as identified by virtue of this one-to-one correspondence, that is, to feel free to replace one set by the other when it is convenient.

1.8.8. If f is a function and A and B are sets, then it can be proved that $f[A \cup B] = f[A] \cup f[B]$ and that $f[A \cap B] \subseteq f[A] \cap f[B]$. The inclusion relation in the case of $A \cap B$ cannot be strengthened.

In elementary mathematics one has occasion to use functions of several variables. Within the framework of our discussion a function of n variables $(n \geq 2)$ is simply a function whose arguments are ordered n-tuples. We can include the case $n = 1$ if we agree that a 1-tuple, $\langle x\rangle$, is simply x. Introducing the notation X^n for the set of all n-tuples $\langle x_1, x_2, \ldots, x_n\rangle$, where each x is a member of the set X, a function, whose domain is X^n and whose range is included in X, is an n-**ary operation in** X. In place of "1-ary" we shall say "unary"; for example, complementation is a unary operation in a power set. In place of "2-ary" we shall say "binary." This was anticipated in our discussion of operations for sets; for example, intersection is a binary operation in a suitable collection of sets. Also, addition in $\mathbb{Z}$ is a binary operation; if $x, y \in \mathbb{Z}$, the value of this function at $\langle x,y \rangle$ is written $x + y$.

The lambda notation may be used for a function of n variables $x_1, \ldots, x_n$, writing $\lambda x_1 \ldots x_n$ in place of λx. For example, a function f of two variables may be denoted by

$$\lambda xy[f(x,y)].$$

Specifically, this denotes a certain function having x as first and y as second variable. In contrast, $\lambda yx[f(x,y)]$ denotes the function with y as first and x as second variable. As a consequence of this convention,

$$\lambda yx[f(x,y)](a,b) = f(b,a)$$

and

$$\lambda xy[f(x,y)] = \lambda yx[f(y,x)].$$

Exercises

1.8.1. Give an example of a function on $\mathbb{R}$ onto $\mathbb{Z}$.

1.8.2. Show that if $A \subseteq X$, then $i_X|A = i_A$.

1.8.3. If X and Y are sets of n and m element, respectively, Y^X has how many elements? How many members of $\mathcal{P}(X \times Y)$ are functions?

1.8.4. Using only mappings of the form $f: \mathbb{Z}^+ \longrightarrow \mathbb{Z}^+$, give an example of a function which

    (a) is injective but not surjective;

    (b) is surjective but not injective.

1.8.5. Let $A = \{1, 2, \ldots, n\}$. Prove that if a map $f: A \longrightarrow A$ is onto, then it is one-to-one, and that if a map $g: A \longrightarrow A$ is one-to-one, then it is onto.

1.8.6. Let $f: \mathbb{R}^+ \longrightarrow \mathbb{R}$, where $f(x) = \int_1^x \frac{dt}{t}$. Show as best you can that f is a bijection.

1.8.7. Prove that the function f defined in Example 1.8.7 is a one-to-one correspondence between $\mathcal{P}(X)$ and 2^X.

1.8.8. Referring to Example 1.8.8, prove that if f is a function and A and B are sets, then $f[A \cup B] = f[A] \cup f[B]$.

1.8.9. Referring to the preceding exercise, prove further that $f[A \cap B] \subseteq f[A] \cap f[B]$, and show that proper inclusion can occur.

1.8.10. Prove that a function f is one-to-one iff for all sets A and B, $f[A \cap B] = f[A] \cap f[B]$.

1.8.11. Prove that a function $f: X \longrightarrow Y$ is onto Y iff $f[X - A] \supseteq Y - f[A]$ for all sets A.

1.8.12. Prove that a function $f: X \longrightarrow Y$ is one-to-one and onto iff $f[X - A] = Y - f[A]$ for all sets A.

§1.9. Composition and Inversion for Functions

To motivate our next definition, we consider an example. Let the functions f and g be defined as follows:

$$f: \mathbb{R} \longrightarrow \mathbb{R} \text{ with } f(x) = 2x + 1,$$

$$g: \mathbb{R}^+ \longrightarrow \mathbb{R}^+ \text{ with } g(x) = x^{1/2}.$$

It is a familiar experience to derive from such a pair of functions a function h for which $h(x) = g(f(x))$. Since the domain of g is $\mathbb{R}^+$ by definition, x must be restricted to real numbers such that $2x + 1 > 0$ for $h(x)$ to be defined. That is, combining f and g in this way yields a function whose domain is the set of real numbers greater than $-\frac{1}{2}$ and whose value at x is $g(f(x)) = (2x + 1)^{1/2}$.

The basic idea of this example is incorporated in the following definition. By using ordered-pair notation (instead of the domain and value notation for functions), we avoid having to make any restriction stemming from a difference between the range of f and the domain of g. The **composite** of functions f and g, symbolized $g \circ f$, is

$$\{\langle x, y \rangle | \text{ there is a } z \text{ such that } xfz \text{ and } zgy\}.$$

It is left to the reader to prove that this relation is a function. This operation for functions is called (functional) **composition**. The following special case of our definition is worthy of note. If $f: X \longrightarrow Y$ and $g: Y \longrightarrow Z$, then $g \circ f: X \longrightarrow Z$ and $(g \circ f)(x) = g(f(x))$.

The reader should recognize that the composite $g \circ f$ of the functions f and g is simply the relative product $f|g$ of the relations f and g. The notation "$g \circ f$" with "f" written on the right for the relative product $f|g$ of two functions leads to consistency with our notation for values of a function (that is, writing "$F(x)$" rather than "xF" for the value of a function F at x).

The above example establishes the fact that functional composition is not a commutative operation; indeed, rarely does $f \circ g = g \circ f$. However, composition is an associative operation. That is, if f, g, and h are functions, then

$$f \circ (g \circ h) = (f \circ g) \circ h.$$

To prove this, assume that $\langle x, u \rangle \in f \circ (g \circ h)$. Then there exists a z such that $\langle x, z \rangle \in g \circ h$ and $\langle z, u \rangle \in f$. Since $\langle x, z \rangle \in g \circ h$, there exists a y such that $\langle x, y \rangle \in h$ and $\langle y, z \rangle \in g$. Now $\langle y, z \rangle \in g$ and $\langle z, u \rangle \in f$ imply that $\langle y, u \rangle \in f \circ g$. Further, $\langle x, y \rangle \in h$ and $\langle y, u \rangle \in f \circ g$ imply that $\langle x, u \rangle \in$

$(f \circ g) \circ h$. Reversing the foregoing steps yields the reverse inclusion and hence equality.

The foregoing proof will be less opaque to the reader if he rewrites it in terms of function values. The proof given is in accordance with our definition of functional composition and has the merit that it avoids any complications arising from a difference between the range of f and the domain of g. From the associative law for composition follows the general associative law, which the reader may formulate. The unique function which is defined by composition from the functions $f_1, f_2, \ldots, f_n$ in that order will be designated by

$$f_1 \circ f_2 \circ \cdots \circ f_n.$$

Examples

1.9.1. Let $h: \mathbb{R} \longrightarrow \mathbb{R}^+$ where $h(x) = (1 + x^2)^{1/2}$. Then $h = g \circ f$ if $f: \mathbb{R} \longrightarrow \mathbb{R}^+$ with $f(x) = 1 + x^2$, and $g: \mathbb{R}^+ \longrightarrow \mathbb{R}^+$ with $g(x) = x^{1/2}$. It is this decomposition of h which is used in computing its derivative.

1.9.2. A decomposition of an arbitrary function along somewhat different lines than that suggested by the preceding example can be given in terms of concepts we have discussed. First we make a definition. If ρ is an equivalence relation with domain X, then

$$j: X \longrightarrow X/\rho \text{ with } j(x) = [x]$$

is onto the quotient set X/ρ; j is the **canonical** or **natural mapping** on X onto X/ρ. Now, if f is a mapping on X into Y, the relation defined by

$$x_1 \rho x_2 \text{ iff } f(x_1) = f(x_2)$$

is clearly an equivalence relation on X. Let j be the canonical map on X onto X/ρ. We contend that a function g on X/ρ into $f[X]$, the range of f, is defined by setting $g[x] = f(x)$. To prove that g is a function, it must be shown that if $[x] = [y]$ then $f(x) = f(y)$. But $[x] = [y]$ iff $x \rho y$ iff $f(x) = f(y)$; so g is a function. Finally, we let i be the inclusion map of $f[X]$ into Y. Collectively, we have defined three functions j, g, i, where

$$j: X \longrightarrow X/\rho \text{ with } j(x) = [x];$$
$$g: X/\rho \longrightarrow f[X] \text{ with } g[x] = f(x);$$
$$i: f[X] \longrightarrow Y \text{ with } i(y) = y.$$

Clearly, j is surjective and i is injective. It is left as an exercise to show that g is bijective and that

$$f = i \circ g \circ j.$$

This equation is the whole point of the discussion. It proves to be a useful decomposition for an arbitrary function f.

1.9.3. In some parts of mathematics a slightly different formalization of the function concept is used. The word "map" is commonly used for the notion we wish to discuss. For this example only we define a **map** to be an ordered pair $\langle f, Y \rangle$ where f is a function and Y is a set, called the **codomain** of the map, that includes R_f. The definitions we have made pertaining to functions require some modifications to be applied to maps. Suppose ϕ is the map $\langle f, Y \rangle$. By definition, the domain D_ϕ of ϕ is equal to D_f. When $x \in D_\phi$, $\phi(x)$ denotes $f(x)$. For the map ϕ the notation $\phi: X \longrightarrow Y$ means that $D_\phi = X$, and Y is the codomain of ϕ. One calls ϕ injective iff f is injective. The terms "surjective" and "bijective" now apply to maps; for example, to say that ϕ is surjective means that R_f is equal to the codomain of ϕ. Suppose that $\psi = \langle g, Z \rangle$ is also a map. By decree, the composite $\psi \circ \phi$ of ϕ and ψ may be formed only if $Y = D_\psi$, and then $\psi \circ \phi$ denotes the map $\langle g \circ f, Z \rangle$. Thus, the domain of $\psi \circ \phi$ is the domain of ϕ and its codomain is the codomain of ψ. Within this framework it is possible to characterize surjectivity and injectivity as properties of maps. The characterization of injectivity that we can demonstrate is the following.

(I) A map $\phi = \langle f, Y \rangle$ is injective iff for each pair of maps $\psi = \langle g, D_f \rangle$ and $\pi = \langle h, D_f \rangle$ having the same domain and D_f as codomain, $\phi \circ \psi = \phi \circ \pi$ implies $\psi = \pi$. Indeed, suppose that ϕ is injective and that Z is the common domain of ϕ and π. Then $\phi \circ \psi = \phi \circ \pi$ implies $\phi(\psi(z)) = \phi(\pi(z))$ for all z in Z. Hence $\psi(z) = \pi(z)$ for all z in Z. Hence $\psi = \pi$. The proof of the converse is left as an exercise.

A characterization of surjectivity is given by a simple alteration of (I).

(II) A map $\phi = \langle f, Y \rangle$ is surjective iff for each pair of maps $\psi = \langle g, Z \rangle$ and $\pi = \langle h, Z \rangle$ having the same codomain and the codomain of ϕ as domain, $\psi \circ \phi = \pi \circ \phi$ implies $\psi = \pi$. The proof is left as an exercise.

With the above characterization at our disposal, the decomposition obtained in Example 1.9.2 can be described for maps as follows. For any map ϕ there exists an injective map i, a surjective map σ, and a bijective map β such that $\phi = i \circ \beta \circ \sigma$.

The inverse, f^{-1}, of a function f is a relation which may not be a function. Indeed, f^{-1} is a function iff $\langle y, x \rangle$ and $\langle y, z \rangle$ in f^{-1} imply that $x = z$. In terms of f this means that if $\langle x, y \rangle$ and $\langle z, y \rangle$ are in f, then $x = z$; that is, f is injective. Thus, f^{-1} is a function iff f is injective. If f is injective, then f^{-1} is called the **inverse function** of f, and f is said to be **invertible**. This operation for one-to-one functions is called (functional) **inversion**. The domain of the inverse function, f^{-1}, of f is the range of f, its range is the domain of f, and $x = f^{-1}(y)$ iff $y = f(x)$. Further, f^{-1} is one-to-one and its inverse, $(f^{-1})^{-1}$, is equal to f. If f is a one-to-one function on X onto Y, then f^{-1} is a one-to-one function on Y onto X. Moreover,

$$f^{-1} \circ f = i_X, \text{ and } f \circ f^{-1} = i_Y.$$

There is another important connection between composition and inversion of functions. If f and g are both injections, then $g \circ f$ is injective, and

$$(g \circ f)^{-1} = f^{-1} \circ g^{-1}.$$

The proof is left as an exercise.

Examples

1.9.4. The function $f: \mathbb{R} \longrightarrow \mathbb{R}$ such that $f(x) = 2x + 1$ is one-to-one. The inverse of f may be written $\{\langle 2x + 1, x\rangle | x \in \mathbb{R}\}$. This is not very satisfying to one who prefers to have a function defined in terms of its domain and its value at each member of the domain. To satisfy this preference, we note that

$$\{\langle 2x + 1, x\rangle | x \in \mathbb{R}\} = \{\langle t, \tfrac{1}{2}(t - 1)\rangle | t \in \mathbb{R}\}.$$

Thus f^{-1} is the function on $\mathbb{R}$ into $\mathbb{R}$ such that $f^{-1}(x) = \tfrac{1}{2}(x - 1)$.

1.9.5. The function $g: \mathbb{R}^+ \longrightarrow \mathbb{R}^+$ such that $g(x) = x^2$ is injective, since $x_1^2 = x_2^2$, and both x_1 and x_2 positive imply that $x_1 = x_2$. Then

$$g^{-1}: \mathbb{R}^+ \longrightarrow \mathbb{R}^+ \text{ where } g^{-1}(x) = x^{1/2}.$$

1.9.6. The function

$$f: \mathbb{R} \longrightarrow \mathbb{R}^+ \text{ where } f(x) = 10^x$$

is known to be bijective. The inverse function is called the **logarithm function** to the base 10, and its value at x is written $\log_{10} x$. The equations

$$\log_{10} 10^x = x, \text{ for } x \in \mathbb{R}, \text{ and } 10^{\log_{10} x} = x, \text{ for } x > 0,$$

are instances of equations $(f^{-1} \circ f)(x) = x$, for $x \in D_f$, and $(f \circ f^{-1})(x) = x$, for $x \in R_f$, which are true for any one-to-one function.

1.9.7. If the inverse of a function f in $\mathbb{R}$ exists, then the graph of f^{-1} may be obtained from that of f by reflection in the line $y = x$. The proof is left as an exercise.

1.9.8. From Example 1.8.8, if the inverse of a function f is defined, then $f^{-1}[A \cup B] = f^{-1}[A] \cup f^{-1}[B]$ and $f^{-1}[A \cup B] \subseteq f^{-1}[A] \cap f^{-1}[B]$. The latter identity can be sharpened to $f^{-1}[A \cap B] = f^{-1}[A] \cap f^{-1}[B]$ for inverse functions. The proof is left as an exercise. A set of the form $f^{-1}[A]$ we call the **inverse** or **counter image** of A under f.

Exercises

1.9.1. Let $f: \mathbb{R} \longrightarrow \mathbb{R}$ where $f(x) = (1 + (1 - x)^{1/3})^{1/5}$. Express f as the composite of four functions, none of which is the identity function.

1.9.2. If $f: X \longrightarrow Y$ and $A \subseteq X$, show that $f|A = f \circ i_A$.

1.9.3. Complete the proof of the assertions made in Example 1.9.2.

1.9.4. Complete the proof of (I) and supply a proof of (II) in Example 1.9.3.

1.9.5. Prove that $f: A \longrightarrow B$ is a bijection iff there exists a map $g: B \longrightarrow A$ such that $g \circ f = i_A$ and $f \circ g = i_B$.

1.9.6. If $f: A \longrightarrow B$ and $g: B \longrightarrow C$ are both bijective, show that $g \circ f: A \longrightarrow C$ is bijective and $(g \circ f)^{-1} = f^{-1} \circ g^{-1}$.

1.9.7. Let $\mathcal{C}$ be a collection of sets. Show that the relation $\sim$ in $\mathcal{C}$, defined by $A \sim B$ iff there exists a bijection between them, is an equivalence relation on $\mathcal{C}$.

1.9.8. For a function $f: A \longrightarrow A$, f^n is the standard abbreviation for $f \circ f \circ \cdots \circ f$ with n occurrences of f. Suppose that $f^n = i_A$. Show that f is one-to-one and onto.

1.9.9. Justify the following restatement of Theorem 1.7.1. Let X be a set. Then there exists a one-to-one correspondence between the equivalence relations on X and the partitions of X.

1.9.10. Prove that if the inverse of the function f in $\mathbb{R}$ exists, then the graph of f^{-1} may be obtained from that of f by a reflection in the line $y = x$.

1.9.11. Show that each of the following functions has an inverse. Determine the domain of each inverse and its value at each member of its domain. Further, sketch the graph of each inverse.

(a) $x \mapsto 2x - 1$, $x \in \mathbb{R}$.

(b) $x \mapsto x^3$, $x \in \mathbb{R}$.

(c) $f = \{\langle x, (1 - x^2)^{1/2}\rangle | 0 \leq x \leq 1\}$.

(d) $f = \left\{ \langle x, \dfrac{x}{x - 1}\rangle | -2 \leq x < 1 \right\}$.

1.9.12. Establish the identity $(g \circ f)^{-1} = f^{-1} \circ g^{-1}$ for one-to-one functions f and g.

1.9.13. Prove that if the inverse of f exists, then $f^{-1}[A \cap B] = f^{-1}[A] \cap f^{-1}[B]$.

1.9.14. The definition of the composite of two relations is applicable to any pair of relations. With this in mind, show that if f is any function and $g = \{\langle y,x\rangle | \langle x,y\rangle \in f\}$ then $g \circ f$ is an equivalence relation.

1.9.15. Let A, B, A', and B' be sets such that A and A' are in one-to-one correspondence and B and B' are in one-to-one correspondence. Show that

(a) there exists a one-to-one correspondence between $A \times B$ and $A' \times B'$;

(b) there exists a one-to-one correspondence between A^B and $A'^{B'}$;

(c) if, further, $A \cap B = \emptyset$ and $A' \cap B' = \emptyset$, then there exists a one-to-one correspondence between $A \cup B$ and $A' \cup B'$.

1.9.16. For sets A, B, and C show that

(a) $A \times B$ is in one-to-one correspondence with $B \times A$;

(b) $(A \times B) \times C$ is in one-to-one correspondence with $A \times (B \times C)$;

(c) $A \times (B \cup C)$ is in one-to-one correspondence with $(A \times B) \cup (A \times C)$.

1.9.17. For sets A, B, and C show that
  (a) $(A \times B)^C$ is in one-to-one correspondence with $A^C \times B^C$;
  (b) $(A^B)^C$ is in one-to-one correspondence with $A^{B \times C}$;
  (c) if, further, $B \cap C = \varnothing$, then $A^{B \cup C}$ is in one-to-one correspondence with $A^B \times A^C$.

§1.10. The Axiom of Choice*

We complete our presentation of the basis for intuitive set theory with the introduction of a third principle.

The axiom of choice. *If $\mathcal{A}$ is a pairwise disjoint collection of nonempty sets, then there exists a set B which consists of one and only one element from each set in $\mathcal{A}$.*

To illustrate the need for such a principle, consider the following situation. The relation $\sim$ in $\mathbb{R}$ defined by $x \sim y$ iff $x - y \in \mathbb{Q}$ is easily seen to be an equivalence relation on $\mathbb{R}$. Hence it determines a partition $\mathcal{P}$ of $\mathbb{R}$. Suppose we wish to define what might be called a "choice set" for this collection $\mathcal{P}$ of pairwise disjoint nonempty sets, that is, a set B consisting of exactly one member from each equivalence class. There appears to be no rule which one could give in advance for choosing a particular element from each member set of $\mathcal{P}$. In other words, there appears to be no predicate form $P(x)$ such that $B = \{x | P(x)\}$. In the absence of such a $P(x)$, the axiom of choice assures us of the existence of a set B having the required property.

To point up the role of the axiom of choice, we consider next a contrasting example. It is possible to devise a partition $\mathcal{P}$ of $\mathbb{N}$ having infinitely many members and such that each member consists of infinitely many natural numbers. The axiom of choice is not needed to assure us of the existence of a choice set for this collection of pairwise disjoint nonempty sets since

$$\{x | \text{For } A \in \mathcal{P}, \ x \text{ is the least member of } A\}$$

is such a set.

A second version of the axiom of choice reads as follows. *If $\mathcal{A}$ is any collection of nonempty sets, then there exists a function, f, with domain $\mathcal{A}$, such that for every $A \in \mathcal{A}$, $f(A) \in A$.* Such a function is called a **choice function** for $\mathcal{A}$. It is obvious that this second version implies the first. Conversely, assume that the first version holds and let $\mathcal{A}$ be any collection

---

* This section may be omitted without loss of continuity. The only essential use we make of this axiom is in §4.6. There we employ a set-theoretic principle known as Zorn's lemma, which is equivalent to the axiom of choice.

of nonempty sets. Then a choice set exists for $\mathcal{B} = \{\{A\} \times A | A \in \mathcal{a}\}$.
From such a set one can easily define a function of the required kind.

We turn next to an application of the axiom of choice that is more directly related to our earlier discussion of functions. It is an easy matter to prove that if $f\colon X \longrightarrow Y$, where X is nonempty, then the following statements are equivalent.

(4) f is injective.

(5) There exists a map $g\colon Y \longrightarrow X$ such that $g \circ f = i_X$.

For proof, let us first deduce (5) from (4). Assume that f is injective and let $Z = R_f$. Then the relation f^{-1} is a function whose domain is Z and whose range is X. If $x \in X$, then $f^{-1}(f(x)) = x$. If $Z = Y$, then f^{-1} is the function we seek. If $Z \subset Y$, then we choose g to be an extension of f^{-1} to Y; for instance, if $x_0 \in X$, then let g be that function on Y into X which agrees with f^{-1} on Z and takes the value x_0 for each $y \in Y - Z$. The proof that (5) implies (4) is just the following calculation. Let $x, x' \in X$ with $f(x) = f(x')$. Then

$$x = i_X(x) = (g \circ f)(x) = g(f(x)) = g(f(x'))$$
$$= (g \circ f)(x') = i_X(x') = x'.$$

A mapping $g\colon Y \longrightarrow X$ satisfying (5) is called a **left inverse** of f. When the equivalence of (4) and (5) is compared with (I) in Example 1.9.3, the characterization (II) of surjectivity in that example might lead one to conjecture the equivalence of the following two statements about a function $f\colon X \longrightarrow Y$.

(6) f is surjective.

(7) There exists a map $h\colon Y \longrightarrow X$ such that $f \circ h = i_Y$.

A map such as h is called a **right inverse** of f. It is easily shown that (7) implies (6). Indeed, assume (7) and let $y \in Y$. Then

$$y = i_Y(y) = (f \circ h)(y) = f(h(y)),$$

so that $y \in R_f$. Hence $Y \subseteq R_f$. But also $R_f \subseteq Y$ and so $Y = R_f$.

The converse is a different matter. Assume that $f\colon X \longrightarrow Y$ is surjective. The condition that a function $h\colon Y \longrightarrow X$ be a right inverse of f is equivalent to

$$h(y) \in f^{-1}[\{y\}].$$

So to construct a right inverse of f we need only "choose" for each $y \in Y$ some element of $f^{-1}\{[y]\}$ and define $h(y)$ to be that element. Although for each $y \in Y$, $f^{-1}[\{y\}]$ is nonempty (because f is surjective), it need not be a unit set; hence, there is no recipe for selecting one element from each of the sets $f^{-1}[\{y\}]$ with $y \in Y$. The axiom of choice is our salvation! Indeed, let

$$\mathcal{C} = \{f^{-1}[\{y\}] | y \in Y\}.$$

By the axiom of choice there exists a choice function c for $\mathcal{C}$. Define $h: Y \longrightarrow X$ by

$$h(y) = c(f^{-1}[\{y\}]).$$

Then, if $y \in Y$, $h(y) \in f^{-1}[\{y\}]$ and $f(h(y)) = y$. Hence $f \circ h = i_Y$.

That it is essential to use the axiom of choice in the above proof follows from the fact that the axiom of choice can be deduced from the assumption that every surjective mapping has a right inverse. Thus, the axiom of choice and the statement that if a function is surjective then it has a right inverse are equivalent.*

The axiom of choice has an interesting history; a full account can be found in Chapter II of Fraenkel and Bar-Hillel (1958), supplemented by Cohen (1966). We shall only remark that apparently the first specific reference to it was made in 1890 in a paper by G. Peano (1858–1932), although others, including Cantor, had used the principle earlier without mention. In 1904 Zermelo proved that the axiom of choice implies that every set can be well-ordered (see §1.11). Cantor had long conjectured that every set can be well-ordered, and a discovery of a proof of this statement was regarded as essential for mathematics. Zermelo's proof, using the axiom of choice, thereby put the axiom of choice in the limelight, and there commenced the controversy about it that only recently has subsided.

§1.11. Ordering Relations

In this section we define several types of relations which have their origin in the intuitive notion of an ordering relation (order of precedence), that is, a relation ρ such that for an appropriate set X there are various distinct members x and y of X such that $x\rho y$, but it is not the case that $y\rho x$. Then, by means of ρ, we could decide to put the x and y in question in the order x,y rather than y,x because $x\rho y$, and it is not the case that $y\rho x$. For a set of real numbers the familiar relations $<$, $\leq$, and $>$ are used in this capacity. For a collection of sets the relations $\subset$ and $\subseteq$ serve similarly.

---

*In Rubin and Rubin (1963) appear a wide variety of statements which are equivalent to the axiom of choice.

The first ordering relation we shall consider has as its defining properties the basic features common to the above relations of $\leq$ for numbers and $\subseteq$ for sets. As a preliminary, we agree to call a relation ρ **antisymmetric** iff for all x and y, if $x\rho y$ and $y\rho x$ then $x = y$. A relation ρ is a **partial ordering** iff it is reflexive, antisymmetric, and transitive. A relation ρ **partially orders** a set X iff $\rho \cap (X \times X)$ is a partial ordering with field X. We note that if ρ is a relation, X is a set, and $\rho' = \rho \cap (X \times X)$, then $F_{\rho'} = F_\rho \cap X$. The relation ρ' is the "restriction" of ρ to X in the sense that it consists of just those members of ρ having both coordinates in X.

If ρ partially orders X, then, according to the definitions above, $\rho' = \rho \cap (X \times X)$ is reflexive on X, antisymmetric, and transitive. Hence, if ρ partially orders X, then $x\rho x$ for x in X, $x\rho y$ and $y\rho x$ with x, y in X imply $x = y$, and $x\rho y$ and $y\rho z$ with x, y, z in X imply $x\rho z$. If ρ is a partial ordering and X is a set, then $\rho \cap (X \times X)$ is a partial ordering. Further, a partial ordering ρ partially orders its field [since $\rho \cap (F_\rho \times F_\rho) = \rho$] and, consequently, any subset of ρ. A relation that is not a partial ordering may partially order some set. For example,

$$\rho = \{\langle X,Y\rangle | X \text{ and } Y \text{ are sets and } X \in Y \text{ or } X = Y\}$$

is not a partial ordering: it is not transitive since $\varnothing \in \{\varnothing\}$ and $\{\varnothing\} \in \{\{\varnothing\}\}$, yet $\varnothing \not\subseteq \{\{\varnothing\}\}$; however, it does partially order $\{0, \{\varnothing\}, \{\varnothing, \{\varnothing\}\}\}$.

Examples

1.11.1. The relation "is an integral multiple of" in $\mathbb{Z}^+$ is a partial ordering.

1.11.2. A hierarchy or a table of organization in a business firm is determined by a partial ordering in some set of positions.

1.11.3. The relation $\rho = \{\langle a,b\rangle, \langle a,c\rangle, \langle c,d\rangle, \langle a,d\rangle, \langle a,a\rangle, \langle b,b\rangle, \langle c,c\rangle, \langle d,d\rangle\}$ is a partial ordering. It partially orders $\{a,b,c\}$, but does not partially order $\{a,b,c,e\}$.

1.11.4. If ρ is a partial ordering, so is ρ^{-1}.

1.11.5. A relation ρ that is reflexive and transitive is a **preordering**. A potential shortcoming of such a relation, in connection with establishing an order of precedence in a set X, is the possibility of ρ being "indifferent" to some distinct pair x, y of objects in the sense that both $x\rho y$ and $y\rho x$. For example, in some population let w be the weight function and h be the height function of individuals, so that $w(x)$ and $h(x)$ are the weight and height, respectively, of the individual named x. Then the relation ρ such that $x\rho y$ iff $w(x) \leq w(y)$ and $h(x) \leq h(y)$ is a preordering, but it does not partially order the population if there are two individuals having the same weight and height.

If ρ is a preordering, then it determines a partial ordering having a partition of $F_\rho\ (= D_\rho)$ as its field according to Exercise 1.7.10. There it is asserted first that the relation $\sim$ such that $x \sim y$ iff $z\rho y$ and $y\rho x$ is an equivalence relation on F_ρ. Secondly, it is stated that the relation ρ' such that $[x]\rho'[y]$ iff $x\rho y$ is a partial ordering having the accompanying set of equivalence classes $[x]$ as field. In summary, from a preordering ρ can be constructed a partial ordering whose field is the set obtained from F_ρ by identifying elements to which ρ is indifferent, or, from a relation ρ which preorders a set A (the reader can supply the definition needed) can be constructed a relation which partially orders the set that results from A upon identification of elements to which ρ is indifferent.

The foregoing is nicely illustrated by taking ρ as the relation in $\mathbb{C}$ such that $z\rho w$ iff the real part of z is less than, or equal to, the real part of w.

We shall follow custom and designate a partial ordering by the symbol $\leq$, which suggests an ordering. We shall use the symbol without the bottom line, that is, $<$, to denote the relation

$$\{\langle x,y\rangle | x \leq y \quad \text{and} \quad x \neq y\}.$$

If $x < y$ we say that x is **less than** y or x **precedes** y or y is **greater than** x. If $\leq$ partially orders a set X, and it is not the case that $x \leq y$ for $x,y \in X$, we write $x \nleq y$. We shall also use $y \geq x$ and $y > x$ as alternatives for $x \leq y$ and $x < y$, respectively, when convenient.

A relation ρ is called **irreflexive** iff for no x in F_ρ is $x\rho x$. Clearly, if $\leq$ is a partial ordering, then $<$ is irreflexive and transitive. Relations which are irreflexive and transitive have a close relationship to partial orderings. Given a set X, let us assign to each partial ordering, $\leq$, with field X the irreflexive and transitive relation $<$ with field $\subseteq X$. This assignment defines a function, f, let us say, on $\mathcal{P} = \{\rho | \rho$ is a partial ordering with field $X\}$ into $\mathcal{Q} = \{\rho | \rho$ is irreflexive and transitive, and $F_\rho \subseteq X\}$. Conversely, let us assign to an irreflexive and transitive relation ρ with $F_\rho \subseteq X$ the relation $g(\rho) = \rho \cup i_X$. Then $g(\rho)$ is a partial ordering with field X. Further, $f \circ g = i_{\mathcal{Q}}$ and $g \circ f = i_{\mathcal{P}}$, so that f is a bijection according to Exercise 1.9.6. Hence f is a one-to-one correspondence between partial orderings with field X and irreflexive and transitive relations with field $\subseteq X$. The derivation of $<$ from $\leq$, and vice versa, can be illustrated in concrete terms by the definition of proper inclusion for sets in terms of inclusion, and vice versa. If $\leq$ partially orders the finite set X, the relation $<$ can be expressed in terms of the following concept. An element y of X is a **cover** of x in X iff $x < y$ and there exists no u in X such that $x < u < y$. If $x < y$, then, clearly, elements $x_1, x_2, \ldots, x_n$ of X can be found such that $x = x_1 <$

$x_2 < \cdots < x_n = y$, and each x_{i+1} covers x_i. Conversely, the existence of such a sequence implies that $x < y$.

A relation ρ is a **simple** (or **linear**) **ordering** iff it is a partial ordering and for all $x, y \in F_\rho$, either $x\rho y$ or $y\rho x$. A relation ρ **simply orders** a set Y iff $\rho \cap (Y \times Y)$ is a simple ordering. The familiar ordering of the real numbers is a typical example of a simple ordering. In contrast, inclusion for sets is not, in general, a simple ordering.

To point out the obvious, the applications of ordering relations are concerned with the determinations of orderings in various sets. In practice, ordering relations for a given set X are usually generated by assigned or proven structural features of X. That is, certain features of X, such as the existence of a particular type of operation or mapping property, will permit the definition of an ordering relation for X; an example of this nature appears in the exercises for this section. Properties of this ordering relation may then prove useful in deducing and describing further features of X. Therefore, it is convenient to have available terminology which gives primary emphasis to the set rather than to an ordering relation for it.

A **partially ordered set** is an ordered pair $\langle X, \leq \rangle$ such that $\leq$ partially orders X. A **simply ordered set** or **chain** is an ordered pair $\langle X, \leq \rangle$ such that $\leq$ simply orders X. For example, if $\mathfrak{F}$ is a collection of sets, then $\langle \mathfrak{F}, \subseteq \rangle$ is a partially ordered set. Again, if $\leq$ is the usual ordering for the integers, then $\langle \mathbb{Z}, \leq \rangle$ is a chain. From the standpoint of set theory, it is more economical to treat ordering relations than ordered sets, that is, sets with accompanying order relations. For example, if $\langle X, \leq \rangle$ is a partially ordered set, then $\leq \cap (X \times X)$ is a partial ordering relation. Instead of dealing with X and a relation $\leq$ which partially orders it, we can deal exclusively with the ordering relation $\leq \cap (X \times X)$, since it determines X as its domain. That is, all statements about ordered sets are equivalent to statements about their ordering relations, and vice versa.

As an illustration of the preceding remark, we restate our earlier characterization of $<$ for a finite set X partially ordered by a relation $\leq$. If $\langle X, \leq \rangle$ is a finite partially ordered set, then $x < y$ iff there exists a chain of the form $x = x_1 < x_2 < \cdots < x_n = y$ in which each x_{i+1} covers x_i. This result enables one to represent any finite partially ordered set by a diagram. The elements of X are represented by dots arranged in accordance with the following rule. The dot for x_2 is placed above that for x_1 iff $x_1 < x_2$, and, if x_2 is a cover of x_1, the dots are joined by a line segment. Thus, $x < y$ iff there exists an ascending broken line con-

necting x with y. Some examples of such diagrams are shown in Figure
1.13. Diagram a is of a chain with five members. Clearly, the diagram

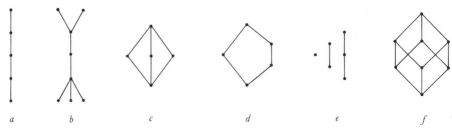

Figure 1.13.

of any chain has this form. Diagram f is of the power set of a three-element set partially ordered by inclusion: the dot at the lowest level
represents the empty subset, the dots at the next level represent the unit
subsets, and so on. Such diagrams not only serve to represent given
partially ordered sets by displaying the ordering relation, but, conversely, also may be used to define partially ordered sets; the ordering
relation is just that indicated by the various broken lines.

In preparation for our next definition concerning partially ordered
sets we discuss an example. The set $\{1,2,3,5,6,10,15,30\}$, whose members are the divisors of 30, is partially ordered by the relation $\leq$ where
$x \leq y$ iff x is a multiple of y. It is left as an exercise to show that the
diagram of this partially ordered set is identical to that given in Figure
1.13 for the subsets of a set of three elements partially ordered by inclusion. Although these two partially ordered sets are obviously not equal,
their structures as partially ordered sets are indistinguishable. This is
the essence of the identity of their respective diagrams. When this type
of relationship exists between two partially ordered sets it is certainly
worthy of note, since any property of one that is expressible in terms of
its ordering relation has an analogue in the other. Thus, we propose to
formalize this type of indiscernibility. The identity of the diagrams of
the two partially ordered sets mentioned above implies, first, the existence of a pairing of the members of the two sets. This can be formulated
as the existence of a one-to-one correspondence, which has the advantage that it does not limit us to finite sets. Next, it is implied that the
relationship between a pair of elements in one set, as specified by the
ordering relation for that set, is the same as that for the corresponding
pair in the other set, relative to its ordering relation. The following

definition is basic in the precise formulation of this property. A function $f: X \longrightarrow X'$ is **order-preserving** (**isotone**) relative to an ordering $\leq$ for X and an ordering $\leq'$ for X' iff $x \leq y$ implies $f(x) \leq' f(y)$. Then the likeness with which we are concerned can be described as the existence of a one-to-one correspondence such that it and its inverse are order-preserving. The customary terminology in this connection follows. An **isomorphism** between the partially ordered sets $\langle X, \leq \rangle$ and $\langle X', \leq' \rangle$ is a one-to-one correspondence between X and X' such that both it and its inverse are order-preserving. If such a correspondence exists, then one partially ordered set is an **isomorphic image** of the other, or, more simply, the two partially ordered sets are **isomorphic.** Thus, the likeness which we observed between the collection of subsets of a three-element set and the set of divisors of 30, with their respective partial orderings, may be expressed by saying that they are isomorphic partially ordered sets.

When the concept of a partially ordered set was defined, it was stated that a collection of sets partially ordered by inclusion is a typical example. This was rather loose talk, since the word "typical" has so many shades of meaning. One precise (and demanding) meaning that might be given is this: Each partially ordered set is isomorphic to a collection of sets partially ordered by inclusion. This is proved next.

THEOREM 1.11.1. A partially ordered set $\langle X, \leq \rangle$ is isomorphic to a collection of sets, indeed, a collection of subsets of X, partially ordered by inclusion.

Proof. For a in X define S_a to be $\{x \in X | x \leq a\}$. Then the mapping f on X into $\{S_a | a \in X\}$ where $f(a) = S_a$ verifies the assertion. The details are left as an exercise.

This result is often stated as: "Each partially ordered set can be represented by a collection of sets (partially ordered by inclusion)." In effect, the theorem means that the study of partially ordered sets is no more general than that of a collection of sets partially ordered by inclusion. In practice the transfer to such a partially ordered set is usually not carried out, since many individual partially ordered sets would lose much of their intuitive content as a result. Finally, we point out that the theorem does *not* assert that each partially ordered set is isomorphic to a collection consisting of all subsets of some set. Such partially ordered sets, that is, those of the form $\langle \mathcal{P}(A), \subseteq \rangle$, do not typify partially ordered sets in general, since they have special features. For example, each contains an element (namely, $\varnothing$) less than every other element and an element (namely, A) greater than every other element.

We conclude this section with the introduction of further terminology for partially ordered sets that will be employed later. A **least** member of a set X relative to a partial ordering $\leq$ is a y in X such that $y \leq x$ for all x in X. If it exists, such an element is unique, so one should speak of *the* least member of X. A **minimal** member of a set X relative to $\leq$ is a y in X such that for no x in X is $x < y$. A minimal member need not be unique, as diagram b in Figure 1.13 illustrates. A **greatest** member of X relative to $\leq$ is a y in X such that $x \leq y$ for all x in X. A greatest element, if it exists, is unique, so one should speak of *the* greatest element of X. A **maximal** member of X relative to $\leq$ is a y in X, such that for no x in X is $x > y$.

A partially ordered set $\langle X, \leq \rangle$ is **well-ordered** iff each nonempty subset has a least member. A familiar example of a well-ordered set is the set of nonnegative integers relative to its natural ordering. Any well-ordered set $\langle X, \leq \rangle$ is a chain, since for two distinct elements x and y of X the set $\{x, y\}$ must have a first element, and hence either $x < y$ or $y < x$.

If $\langle X, \leq \rangle$ is a partially ordered set and $A \subseteq X$, then an element x in X is an **upper bound** for A iff, for all a in A, $a \leq x$. Similarly, an element x in X is a **lower bound** for A iff, for all a in A, $x \leq a$. A set may have many upper bounds. An element x in X is a **least upper bound** or **supremum** for A (symbolized, lub A or sup A) iff x is an upper bound for A and $x \leq y$ for all upper bounds for A. In other words, a supremum is an upper bound which is a lower bound for the set of all upper bounds. An element x in X is a **greatest lower bound** or **infinum** for A (symbolized glb A or inf A) iff x is a lower bound for A and $x \geq y$ for any lower bound y for A. It is immediate that if A has a least upper bound, then it is unique, and that the same is true for a greatest lower bound.

Exercises

1.11.1. Let ρ be a relation having X as its field.
   (a) Show that ρ is a partial ordering iff $i_X \subseteq \rho$, $\rho \cap \rho^{-1} \subseteq i_X$, and $\rho | \rho \subseteq \rho$.
   (b) Use the result in (a) to prove that if ρ is a partial ordering, then so is ρ^{-1}.
1.11.2. For the set of real-valued continuous functions with the nonnegative

reals as domain, define $f = O(g)$ to mean that there exist positive constants M and N such that $f(x) \leq Mg(x)$ for all $x > N$. Show that this is a preordering and define the associated equivalence relation.

1.11.3. Supply proofs of the assertions in the text that culminate with the determination of a one-to-one correspondence between partial orderings with field X and irreflexive and transitive relations with field included in X.

1.11.4. If $\leq$ is a partial ordering, show that $<$ is an irreflexive and transitive relation. Conversely, if $<$ is an irreflexive and transitive relation, show that the relation $\leq$ such that $x \leq y$ iff $x < y$ or $x = y$ is a partial ordering.

1.11.5. For what sets A is $\langle \mathcal{P}(A), \subseteq \rangle$ a simply ordered set?

1.11.6. Let $\langle X, \leq \rangle$ and $\langle X', \leq' \rangle$ be partially ordered sets. Show that $X \times X'$ is partially ordered by ρ where $\langle x, x' \rangle \rho \langle y, y' \rangle$ iff $x \leq y$ and $x' \leq' y'$. The partially ordered set $\langle X \times X', \rho \rangle$ is the (cartesian) **product** of the given partially ordered sets.

1.11.7. The **dual** of a partially ordered set $\langle X, \rho \rangle$ is the partially ordered set $\langle X, \rho^{-1} \rangle$. If $\langle X, \leq \rangle$ is a partially ordered set and $a, b \in X$ with $a \leq b$, then the set of all x in X, such that $a \leq x \leq b$, is called the **closed interval** $[a,b]$. Show that the set of intervals of a partially ordered set $\langle X, \leq \rangle$, partially ordered by inclusion, is isomorphic to a subset of the product of $\langle X, \leq \rangle$ and its dual.

1.11.8. A partially ordered set is **self-dual** if it is isomorphic to its dual. Show that

      (a) there are just two nonisomorphic partially ordered sets of two elements, both of which are self-dual, and

      (b) there are five nonisomorphic partially ordered sets of three elements, three of which are self-dual.

1.11.9. Show by an example that if $\langle X, \leq \rangle$ and $\langle X', \leq' \rangle$ are partially ordered sets and $f: X \longrightarrow X'$ is a one-to-one correspondence which preserves order, then f^{-1} need not preserve order.

1.11.10. Given that f is an isomorphism between the partially ordered sets $\langle X, \leq \rangle$ and $\langle X', \leq' \rangle$, show that $x < y$ iff $f(x) <' f(y)$.

1.11.11. Supply details for the proof of Theorem 1.11.1.

1.11.12. Let $\langle X, \leq \rangle$ be a partially ordered set. Show that u is a maximal element iff $y \in X$ and $y \geq u$ imply $y = u$. Show that v is a minimal element iff $y \in X$ and $y \leq v$ imply $y = v$.

1.11.13. Let $\mathcal{F}_n$ be the collection of all subsets of $\mathbb{Z}^+$ which have at most n members for n a fixed positive integer, and let $\mathcal{F}$ be the collection of all finite subsets of $\mathbb{Z}^+$. Show that, relative to inclusion,

      (a) each element of $\mathcal{F}_n$ having n members is maximal, and

      (b) $\mathcal{F}$ has no maximal elements.

1.11.14. As the elements of a set X we take all square regions which lie inside a given rectangular region which is itself not a square. Relative to inclusion, what are the maximal elements of X?

1.11.15. Show that in a chain the notions of a greatest element and a maximal element coincide, and show the same for a least element and a minimal element.

1.11.16. Let $\langle X, \leq \rangle$ be a partially ordered set with the property that each nonempty subset which has an upper bound has a least upper bound. Show that each nonempty subset of X which has a lower bound has a greatest lower bound.

1.11.17. Show that if $\langle X, \leq \rangle$ is a well-ordered set, then it has the property assumed for the partially ordered set in the preceding exercise.

1.11.18. Let X be a set and ρ an operation in X. (Thus, ρ is a function on $X \times X$ into X; let us denote the value of ρ at $\langle x, y \rangle$ by xy.) Suppose that ρ is commutative, associative, and idempotent [that is, $xy = yx$, $x(yz) = (xy)z$, and $xx = x$ for all $x, y, z \in X$]. For x, $y \in X$ define $x \leq y$ iff $x = xy$. Show that

    (a) $\leq$ partially orders X,

    (b) if X has a least element 0, then $0x = 0$,

    (c) $xy \leq x, y$ and if $z \leq x, y$ then $xy \geq z$.

1.11.19. The relation $\leq$ where $m \leq n$ iff m divides n partially orders $\mathbb{Z}^+$. Show that each pair of integers has a least upper bound and a greatest lower bound relative to this ordering.

1.11.20. Show that each subset of $\mathcal{P}(A)$ partially ordered by inclusion has a least upper bound and a greatest lower bound.

§1.12. Proof and Definition by Induction

Proofs and definitions by induction are commonplace in all branches of mathematics. To develop and justify these techniques requires an understanding of the natural number sequence

$$0, 1, 2, \ldots,$$

where we rely on the dots to suggest the continuation of the sequence beyond the members displayed. We cannot expect that the natural number sequence can be defined in terms of anything essentially more primitive than itself. All we can do is elaborate on what our conception of it is, in terms of notions already discussed. As our initial description, we say that the natural numbers are those objects which can be generated by starting with an initial object 0 (zero) and, from each object already generated, passing to another uniquely determined object n', the successor of n. Moreover, objects differently generated are distinct. Here "n'" may be thought of as an alias for "$n + 1$." The accent notation is used to emphasize that $'$ is a primitive operation used in generating the natural numbers (and, thus, not to be confused with addition, which can be defined later as a binary operation in the set). Applying this operation to an object n already generated produces the object n', which is "next after" or "succeeds" n. The natural numbers thereby appear as the objects

$$0, 0', (0')', ((0')')', \ldots$$

or, more simply,

$$0, 0', 0'', 0''', \ldots.$$

The transition to the usual notation for natural numbers is made upon introducing $1, 2, \ldots, 9$ to stand for $0', 0'', \ldots, 0''''''''''$ and then employing decimal notation.

This description can be dissected into five "atomic" components.

1. 0 is a natural number.
2. If n is a natural number, then n' is a natural number.
3. For all natural numbers m and n, if $m' = n'$ then $m = n$.
4. For all natural numbers n, $n' \neq 0$.
5. The only natural numbers are those given by 1 and 2.

Statements 3 and 4 insure that differently generated natural numbers are distinct. To see this, suppose that at a given stage in the generation $0, 1, \ldots, n$ have been generated and are distinct from each other. Then the next number n' must be distinct from these. Assume, to the contrary, that $n' = m$, where m is a number already generated. Then $m \neq 0$ by 4 and so $m = k'$ for some k already generated. But then, by 3, $n = k$, contrary to the assumed distinctness of $0, 1, \ldots, n$.

The foregoing description of the natural numbers makes no reference to the totality of natural numbers as a completed entity. The existence of a set having properties 1–5 cannot be inferred from any of the general principles of set theory described so far. To cope with these circumstances, we propose a definition and an assumption. By a **Peano system*** we shall mean an ordered triple $\langle D,s,0 \rangle$ consisting of a set D (the domain of the system), a unary operation s in D, and an element 0 of D, for which the following conditions hold.

P1.　For all $x,y \in D$, if $sx = sy$, then $x = y$.
P2.　For all $x \in D$, $sx \neq 0$.
P3.　If C is any subset of D such that (i) $0 \in C$ and (ii) whenever $x \in C$ then $sx \in C$, then $C = D$.

The assumption that we make is: there exists a Peano system.

Let us show that the members of the domain D of a Peano system $\langle D,s,0 \rangle$ enjoy properties 1–5 (upon rewriting them, substituting "member

---

* The reason for this name will be supplied presently.

of D" for "natural number" and "*sn*" for "*n′*") that we have attributed
to the natural numbers. Properties 1 and 2 hold, for they have been
incorporated into the definition of a Peano system. Properties 3 and 4
are restatements of P1 and P2, respectively. To show that 5 holds, we
must demonstrate that if $x \in D$ then either $x = 0$ or there exists a
$y \in D$ such that $x = sy$. Let $A = \{x \in D | x = 0$ or $x = sy$ for some
$y \in D\}$. We shall prove that $A = D$ using P3. Clearly $A \subseteq D$ and
$0 \in A$. Next, assume that $x \in A$; we wish to show that $sx \in A$, that is,
that $sx = 0$ or $sx = sz$ for some $z \in D$. The first possibility is excluded
by P2, but the second is trivially satisfied by choosing z as x. Hence, by
P3, $A = D$.

Turning matters around, our initial description of the system of natural numbers may be summarized by the statement that the natural
number sequence, with ′ and 0, is a Peano system. To what degree does
this statement characterize the natural number sequence as we understand that system intuitively? It is a *complete* characterization to within
the notation used, for it can be proved that any two Peano systems are
isomorphic; that is, if $\langle D, s, 0 \rangle$ and $\langle D', s', 0' \rangle$ are Peano systems, then
there exists a bijection $F: D \longrightarrow D'$ such that $F(0) = 0'$ and, for each
$x \in D$, $F(sx) = s'F(x)$. In other words, the domains of two systems can
be paired so that successors correspond to successors and, thus, whatever is observed, defined, or proved in one can be mimicked in the other.
The fact that, as understood intuitively, the natural numbers form a
Peano system, coupled with the result that any two Peano systems are
isomorphic, lead us to conceive of the natural number sequence as some
one Peano system. We shall denote the domain of this system by $\mathbb{N}$, the
successor operation by ′, and its elements by

$$0, 1, 2, \ldots,$$

where $0' = 1$, $1' = 2$, and so on. Generally we shall identify this system
by simply $\mathbb{N}$. Temporarily let us assume that binary operations $+$ and
$\cdot$ have been defined in $\mathbb{N}$ and that it has been shown that these operations have all of the properties attributed to addition and multiplication,
as understood intuitively. One part of the definition of $+$ is that $n' =
n + 1$, so that we shall feel free to use this fact. Finally, we assume that
a binary relation $\leq$ has been defined in $\mathbb{N}$ and that it has been shown
to have the properties attributed to the familiar ordering relation.

As a Peano system, condition P3 holds for $\mathbb{N}$. We may state it as:
If $M \subseteq \mathbb{N}$ and (i) $0 \in M$ and (ii) whenever $m \in M$ then $m + 1 \in M$,

then $M = \mathbb{N}$. An immediate consequence is the **principle of mathe-matical induction:** Let $P(n)$ express a property of n, where $n \in \mathbb{N}$ If.

(i) $P(0)$ is true, and

(ii) for all natural numbers, $P(m)$ implies $P(m + 1)$,

then $P(n)$ is true for all natural numbers n. A consideration of $M = \{n \in \mathbb{N} | P(n)$ is true$\}$ leads immediately to the conclusion.

A proof which appeals to this principle to establish that some property of natural numbers holds for all natural numbers is called a **proof by induction.** In connection with such proofs we use the following ter-minology. The proof that $P(0)$ is true we call the **induction basis** and the proof that $P(m)$ implies $P(m + 1)$ is called the **induction step.** For the induction step one usually assumes that $P(m)$ is true and then de-duces that $P(m + 1)$ is true. Within the induction step, the assumption that $P(m)$ is true is called the **induction hypothesis.** We present next several illustrations of proof by induction.

Examples

1.12.1. Let $P(n)$ be "$6^{n+2} + 7^{2n+1}$ has 43 as a factor." Then $P(0)$, which is "$6^2 + 7$ has 43 as a factor" is true. Assume $P(m)$ is true. To deduce that $P(m + 1)$ is true we make the following computation.

$$
\begin{aligned}
6^{(m+1)+2} + 7^{2(m+1)+1} &= 6^{m+3} + 7^{2m+3} \\
&= 6 \cdot 6^{m+2} + 49 \cdot 7^{2m+1} \\
&= 6 \cdot 6^{m+2} + (6 + 43) \cdot 7^{2m+1} \\
&= 6 \cdot (6^{m+2} + 7^{2m+1}) + 43 \cdot 7^{2m+1}
\end{aligned}
$$

By the induction hypothesis the first summand to the right of the last equality sign has 43 as a factor and by inspection so does the second. Hence $P(m + 1)$ is true. Thus $P(n)$ is true for all n.

1.12.2. Let $P(n)$ be "If a set has n elements, then its power set has 2^n ele-ments." Since the power set of the empty set has one element, $P(0)$ is true. Assume that $P(m)$ is true and consider a set $A = \{a_1, a_2, \ldots, a_m, a_{m+1}\}$ of $m + 1$ elements. Every subset of $B = \{a_1, a_2, \ldots, a_m\}$ is a subset of A. So is each set that results upon adjoining the element a_{m+1} to a subset of B. The reader is asked to verify that the totality of sets we have just mentioned are distinct from one another and exhaust $\mathcal{P}(A)$. By the induction hypotheses, $\mathcal{P}(B)$ has 2^m ele-ments and, hence, so does the collection of sets obtained by adjoining a_{m+1} to an element of $\mathcal{P}(B)$. Thus, $\mathcal{P}(A)$ has $2^m + 2^m = 2^{m+1}$ elements. Since $P(m + 1)$ is true, it follows that $P(n)$ is true for all n.

1.12.3. A slight modification of the principle of mathematical induction reads: If $P(n_0)$ is true and if $P(m)$ implies $P(m + 1)$ for every natural number m greater than n_0, then $P(n)$ is true for all natural numbers n greater than or equal to n_0. This form of the principle is needed to show, for example, that if $n \geq 4$ then $2^n \leq n!$ A similar modification holds for the variation of the principle of induction that we present below.

The theorem which follows is the basis for a variant form of the principle of mathematical induction.

THEOREM 1.12.1. If M is a subset of $\mathbb{N}$ such that
   (i) $0 \in M$, and
   (ii) whenever a number m and all numbers less than m are in M, then $m + 1 \in M$,
then $M = \mathbb{N}$.

Proof. Suppose M satisfies the given hypotheses. Let

$$M' = \{n \in \mathbb{N} | r \in M \text{ for all } r \leq n\}.$$

Then $0 \in M'$. Assume $n \in M'$. Then for all $r \leq n$, $r \in M$ and, hence, by (ii), $n + 1 \in M$. Let s be a natural number such that $s \leq n + 1$. Then $s \leq n$ or $s = n + 1$ and so $s \in M'$. Thus, for all $s \leq n + 1$, $s \in M$. Hence, $n + 1 \in M'$. It follows that $M' = \mathbb{N}$. Since $M' \subseteq M$ from the definition of M', we conclude that $M = \mathbb{N}$.

From this theorem follows immediately the **principle of complete induction:** Let $P(n)$ express a property of n, where $n \in \mathbb{N}$. If
   (i) $P(0)$ is true, and
   (ii) for each natural number m, $P(r)$ for all $r \leq m$ implies $P(m + 1)$,
then $P(n)$ is true for all natural numbers n.

This is sometimes called the **strong form** of the principle of induction. The justification for this designation follows. In both cases we obtain the conclusion that for all n, $P(n)$ is true. The first case requires that we show that $P(m)$ implies $P(m + 1)$ for all natural numbers m; for the second case we need show that $P(m)$ implies $P(m + 1)$ for only those m's such that $P(r)$ is true for all $r \leq m$. Thus the second case gives the same conclusion as the first with a weaker hypothesis.

Examples

1.12.4. As an illustration of a proof by strong induction, we prove the theorem that every integer greater than 1 has a prime factor, starting the induction with 2. Obviously 2 has a prime factor. Assume the theorem for all m with $2 \leq m \leq n$ and consider $n + 1$. If $n + 1$ has no factor a with $1 < a < n + 1$, then $n + 1$ is a prime and has itself as a prime factor. If $n + 1$ has a factor a with

$1 < a < n + 1$, then $2 \leq a \leq n$. By the induction hypothesis a has a prime factor b, which is then a prime factor of $n + 1$. Thus, in every case, $n + 1$ has a prime factor.

1.12.5. As a somewhat more important illustration of proof by strong induction, we prove next what is often called the **fundamental theorem of arithmetic:** Every natural number greater than 1 has a representation as a product of primes that is unique to within the order of the factors. Again we begin the induction with 2. Clearly, 2 has such a representation. Assume that all numbers less than n have unique representations and consider n. The set of divisors of n which are greater than 1 is nonempty and, hence, has a least member p. Then p is a prime since a divisor q of p with $1 < q < p$ would be a smaller divisor of n. If $n = pn_1$, then n_1 has a unique representation by the induction hypothesis. Replacing n_1 by its unique representation as a product of primes yields a representation of $n = pn_1$ as a product of primes, and this is the only representation of n which contains p as a factor. If the theorem is false for n then it has a second representation. If q is the smallest prime present in this second representation, then $q > p$, since this other representation of n does not involve p and p is the smallest divisor (>1) of n. Let $n = qn_2$ and $q = p + d$. Then $n = pn_2 + dn_2$. Since p divides n, p divides dn_2. Now $dn_2 < n$ and, consequently, has a unique representation. Hence, p divides d or p divides n_2. But p is not a factor of n_2 since it contains no factor less than q and $q > p$. Thus p divides d. Let $d = rp$. Then $q = p + rp = p(1 + r)$. This is a contradiction, since q is a prime. Thus n has no decomposition other than the essentially unique decomposition with p as a factor.

We turn now to applications of induction to defining things instead of proving things. The reader is familiar with the definition of factorial n, symbolized $n!$, by way of the pair of equations

$$0! = 1, \qquad (n + 1)! = n! \cdot (n + 1).$$

The supporting argument that thereby a function on $\mathbb{N}$ into $\mathbb{N}$ is defined runs as follows. The first equation defines the function at 0. The second equation, when evaluated for $n = 0$, gives $1! = 0! \cdot 1 = 1 \cdot 1 = 1$. The second equation, when evaluated for $n = 1$, gives $2! = 1! \cdot 2 = 1 \cdot 2 = 2$. Proceeding in this manner, $n!$ is determined for any given natural number n. Thus a function whose domain is $\mathbb{N}$ is defined. There is an error in this intuitive reasoning. To disclose it we recall that a function is a set, so that to define a function is to define a set of a certain kind. The procedure just employed permits one to define as many members as one chooses (namely, $\langle 0,0! \rangle$, $\langle 1,1! \rangle$, . . . , $\langle n,n! \rangle$ for a preassigned n) of a certain set, but it does not yield a definition of the set consisting of all such ordered pairs. The error in the intuitive reasoning consists in using a function symbol without first giving a function for it to denote.

Since no one doubts that the equations above define a function and, indeed, exactly one function, it should be possible to prove that this is so. The theorem which comes to the rescue is the following.

THEOREM 1.12.2. Let X be a set and c an element of X. Let g be a function on $\mathbb{N} \times X$ into X. Then there exists exactly one function $f: \mathbb{N} \longrightarrow X$ such that

$$f(0) = c \text{ and } f(n + 1) = g(n, f(n)).$$

Using the principle of induction one can easily show that there exists at most one function satisfying the conditions. The proof of the existence of such a function is quite lengthy and will not be given. To illustrate the theorem, take $X = \mathbb{N}$, $c = 1$, and let $g: \mathbb{N} \times \mathbb{N} \longrightarrow \mathbb{N}$ be the map defined by $g(n,m) = m \cdot (n + 1)$. Then, according to the theorem, there exists exactly one function f on $\mathbb{N}$ into $\mathbb{N}$ such that $f(0) = 1$ and $f(n + 1) = g(n, f(n)) = f(n) \cdot (n + 1)$. This establishes the existence and uniqueness of the factorial function.

A function whose existence and uniqueness is established by a theorem having the form of Theorem 1.12.2 is said to be **defined by recursion or recursively** (or **by induction**).

Another example of a definition by induction (as well as two proofs by induction) occurs in the derivation of the general associative law for an associative binary operation in a set. In §1.5 we referred to this result for the case of the union and intersection operations for sets; in §1.9 we referred to it pertaining to the composition of functions. The setting in which to view the general result is as follows. Let X be a set and $f: X^2 \longrightarrow X$ a binary operation in X. We shall write $a * b$ for the image of $\langle a,b \rangle \in X^2$ under f. In terms of f, two ternary operations in X—that is, mappings on X^3 into X—can be defined. One of these maps $\langle a,b,c \rangle$ onto $(a * b) * c$ and the other maps $\langle a,b,c \rangle$ onto $a * (b * c)$. Similarly, five 4-ary operations in X can be defined. More generally, we can define inductively n-ary operations in X. Assume these have been constructed from f to the stage of m-ary operations for every m such that $1 \leq m < n$. It is understood here that the identity map is taken for the case $m = 1$. Now let m be any natural number such that $1 \leq m < n$ and let

$$\langle a_1, a_2, \ldots, a_m \rangle \mapsto u(a_1, a_2, \ldots, a_m),$$

$$\langle a_{m+1}, a_{m+2}, \ldots, a_n \rangle \mapsto v(a_{m+1}, a_{m+2}, \ldots, a_n),$$

be specific m-ary and $(n - m)$-ary operations determined by f. Then the mapping

$$\langle a_1, a_2, \ldots, a_n \rangle \mapsto u(a_1, a_2, \ldots, a_m) * v(a_{m+1}, a_{m+2}, \ldots, a_n)$$

is an n-ary operation. All the maps resulting in this way by varying m, u, and v are the n-ary operations associated with f. The results of applying these mappings to n-tuples $\langle a_1, a_2, \ldots, a_n \rangle$ will be called **composites** of $a_1, a_2, \ldots, a_n$ (in that order). If f is an **associative operation** (or, as some say, satisfies the associative law), that is, for all a, b, and c in X

$$a *(b * c) = (a * b) * c,$$

then the various composites are equal to one another. This is the general associative law which we now prove.

THEOREM 1.12.3. Let $\langle a,b \rangle \mapsto a * b$ be an associative operation in a set X. Then all composites of $a_1, a_2, \ldots, a_n$ (in that order) are equal. The common value will be written $a_1 * a_2 * \cdots * a_n$.

Proof. We first define a particular composite $\pi_1^m a_i$, $m \leq n$, by the formulas

$$\pi_1^1 a_k = a_1, \qquad \pi_1^{r+1} a_k = (\pi_1^r a_k) * a_{r+1}.$$

This is an inductive definition of the type described in Theorem 1.12.1, but with the domain of the function restricted to a segment, $1 \leq m \leq n$, of $\mathbb{N}$. For these composites we prove that

(8) $\pi_1^r a_k * \pi_1^s a_{r+j} = \pi_1^{r+s} a_k$.

The proof is by induction on s. For $s = 1$ this relation is true by definition. Assume it is true for $s = m$, and consider the case where $s = m + 1$. We have

$$\begin{aligned}
\pi_1^r a_k * \pi_1^{m+1} a_{r+j} &= \pi_1^r a_k * ((\pi_1^m a_{r+j}) * a_{r+m+1}) \\
&= ((\pi_1^r a_k) * (\pi_1^m a_{r+j})) * a_{r+m+1} \\
&= (\pi_1^{r+m} a_k) * a_{r+m+1} \\
&= \pi_1^{r+m+1} a_k
\end{aligned}$$

as desired.

This property of the particular composite defined is used to prove next by strong induction on n that any composite of $a_1, a_2, \ldots, a_n$, $n \geq 1$, is equal to $\pi_1^n a_k$. Clearly this is true for $n = 1$. Assume it is true for all composites of r elements of X with $r \leq m$, and consider any composite associated with $\langle a_1, a_2, \ldots, a_{m+1} \rangle$. By definition it is a composite $b * c$, where b is a composite associated with $\langle a_1, a_2, \ldots, a_s \rangle$

and c is a composite associated with $\langle a_{s+1}, a_{s+2}, \ldots, a_{m+1} \rangle$. By the induction hypothesis

$$b = \pi_1^s \, a_k, \qquad c = \pi_1^{m+1-s} \, a_{s+j}.$$

Then $b * c = \pi_1^{m+1} \, a_k$ by (8). It follows that, for all n, all composites of $a_1, a_2, \ldots, a_n$ are equal, each being equal to $\pi_1^n \, a_k$.

We continue our discussion of recursive definitions by singling out a special case of Theorem 1.12.2, which reads as follows.

THEOREM 1.12.4.   Let X be a set and c be an element of X. Let G be a function on X into X. Then there exists exactly one function $f: \mathbb{N} \longrightarrow X$ such that

$$f(0) = c \text{ and } f(n+1) = G(f(x)).$$

This follows from Theorem 1.12.2 upon choosing g to be the function on $\mathbb{N} \times X$ into X such that $g(n,x) = G(x)$. Since Theorem 1.12.2 holds for any Peano system, so does Theorem 1.12.4, which when stated for a Peano system $\langle D,s,0 \rangle$ reads: Let X be a set and c be an element of X. Let G be a function on X into X. Then there exists exactly one function $F: D \longrightarrow X$ such that

$$F(0) = c \text{ and } F(sx) = G(F(x)).$$

With this result can be proved the earlier assertion that two Peano systems, $\langle D,s,0 \rangle$ and $\langle D',s',0' \rangle$, are isomorphic. One begins by choosing D' as X, $0'$ as c, and s' as G. The function F so obtained satisfies the following two properties required of an isomorphism:

$$F(0) = 0' \text{ and } F(sx) = s'F(x).$$

It remains to show that F is a bijection. That F is a surjection and F is an injection can both be proved by induction (see Exercise 1.12.11).

Theorem 1.12.2 has a variety of generalization. That stated next can be used to give recursive definitions of operations in $\mathbb{N}$.

THEOREM 1.12.5.   Let g be a function on $\mathbb{N} \times \mathbb{N}$ into $\mathbb{N}$ and h be a function on $\mathbb{N}$ into $\mathbb{N}$. Then there exists a unique function $f: \mathbb{N} \times \mathbb{N} \longrightarrow \mathbb{N}$ such that

$$f(n,0) = h(n) \text{ and } f(n,m') = g(n,f(n,m)).$$

If we take $h(n) = n$, $g(n,m) = m'$, and denote by $n + m$ the value $f(n,m)$ of the resulting function, then

$$n + 0 = n \text{ and } n + m' = (n + m)'.$$

The uniquely determined operation so defined can be shown to have the properties of addition. If we take $h(n) = 0$, $g(n,m) = m + n$, and denote by $n \cdot m$ the value $f(n,m)$ of the resulting function, then

$$n \cdot 0 = 0 \text{ and } n \cdot m' = n \cdot m + n.$$

The operation so defined can be shown to have the properties of multiplication. In summary, the conception of the natural number sequence $\mathbb{N}$ as a Peano system $\langle \mathbb{N},',0 \rangle$ provides an adequate basis for developing arithmetic.*

We have called a system which satisfies the conditions P1–P3 a Peano system because P1–P3 correspond directly to the axioms for the natural numbers given by G. Peano in his booklet published in 1889 [a translation appears in van Heijenoort (1967)]. It is known, through Peano's own acknowledgment, that Peano borrowed his axioms from Dedekind, who, in his remarkable little book *Was sind und was sollen die Zahlen* [a translation appears in Dedekind (1901)], gave a much more comprehensive account of the derivation of properties of the natural numbers from a set of axioms. This work used the language of set theory and simultaneously contributed to the development of a theory of sets. It also gave a complete account of the theory of natural numbers as a branch of set theory and of proof and definition by induction. Even earlier, Frege carried out a development of the system of natural numbers. In Frege [1879; translated in van Heijenoort (1967)] he offered a theory of sequences which he subsequently used [in Frege (1884)] to define the natural numbers.

Exercises

1.12.1. Prove that for all natural numbers n, $|\sin nx| \le n|\sin x|$.

1.12.2. Prove that if $a > 1$ then, for all natural numbers n, $(1 + a)^n \ge 1 + an$.

1.12.3. Prove that if $n \ge 1$, $\sum_1^n \dfrac{1}{(j + 1)!} = 1 - \dfrac{1}{(n + 1)!}$.

---

* A detailed development of the arithmetic of the natural numbers is given in Fefermar (1964), Chapter 3.

1.12.4. Prove that for all natural numbers n,

$$(\cos u + i \sin u)^n = \cos nu + i \sin nu$$

where $i^2 = -1$.

1.12.5. Prove that if $n \geq 1$ and $\sin u \neq 0$, then

$$\cos u \cdot \cos 2u \cdot \cdots \cdot \cos 2^{n-1}u = \frac{\sin 2^n u}{2^n \sin u}.$$

1.12.6. Prove that for all natural numbers n,

$$(2n)! \leq 2^{2n}(n!)^2.$$

1.12.7. Prove that if $n \geq 1$,

$$D^n(\ln x) = (-1)^{n-1}(n-1)!\,x^{-n}.$$

1.12.8. Prove that if $n \geq 1$, $n^n e^{-n} < n!$.

1.12.9. Given a set of n (≥ 2) points in the plane, no three of which are collinear, prove that the number of straight lines joining these points is $n(n-1)/2$.

1.12.10. Let $*$ be an associative and commutative ($a * b = b * a$ for all a and b) operation in a set X. Prove the general commutative law: If $1', 2', \ldots, n'$ is $1, 2, \ldots, n$ in any order, then

$$a_1 * a_2 * \cdots * a_n = a_{1'} * a_{2'} * \cdots * a_{n'}.$$

1.12.11. Let $\langle D,s,0 \rangle$ and $\langle D',s',0' \rangle$ be Peano systems and F a function on D into D' such that $F(0) = 0'$ and $F(sx) = s'F(x)$. Show that F is a bijection (and, hence, an isomorphism) by proving that

    (a) $\{y \in D' |$ for some $x \in D$, $F(x) = y\} = D'$, and

    (b) $\{x \in D |$ for all $z \in D$, if $F(z) = F(x)$, then $z = x\} = D$.

CHAPTER TWO

Logic

As we shall study it, mathematical or symbolic logic has two aspects. On one hand it is logic—it is an analytical theory of the art of reasoning whose goal is to systematize and codify principles of valid reasoning. It has emerged from a study of the use of language in argument and persuasion, and is based on the identification and examination of those parts of language which are essential for these purposes. It is formal in the sense that it lacks reference to meaning. Thereby it achieves versatility: it may be used to judge the correctness of a chain of reasoning (in particular, a "mathematical proof") solely on the basis of the form (and not the content) of the sequence of statements which make up the chain. There is a variety of symbolic logics. We shall be concerned solely with that one which encompasses most of the deductions of the sort encountered in mathematics. Within the context of logic itself, this is "classical" symbolic logic.

The other aspect of symbolic logic is interlaced with problems relating to the foundations of mathematics. In brief, it amounts to formulating a mathematical theory as a logical system augmented by further axioms; in Chapter 3, which treats axiomatic theories, some indication will be given of this approach to mathematical theories. The idea of regarding a mathematical theory as an "applied" system of logic originated with Frege, who devised a system of logic [in Frege (1879)] for use in his study of the foundations of arithmetic [in Frege (1884)]. Frege's work was continued in Whitehead and Russell (1910–13), which has been described as a demonstration that mathematics can be reduced to logic. This description is not generally accepted, and the claim that such a reduction is possible seems untenable; for instance, the assumption that a Peano system exists (see §1.12) can hardly be claimed to be a principle of logic.

In this chapter, logic is treated semantically: the symbolism that is introduced is interpreted in a meaningful way. An alternative, the syntactical approach, which is based on the manipulation of symbols or strings of symbols in accordance with prescribed rules, is discussed in Chapter 3.

§2.1.   The Statement Calculus. Sentential Connectives

In mathematical discourse and elsewhere, one constantly encounters declarative sentences which have been formed by modifying a sentence with the word *not* or by connecting sentences with the words *and, or, if . . . then* (or *implies*), and *if and only if.* These five words or combinations of words are called **sentential connectives.** Our first concern here is the analysis of the structure of a **composite sentence** (that is, a declarative sentence in which one or more connectives appear) in terms of its constituent **prime sentences** (that is, sentences which either contain no connectives or, by choice, are regarded as "indivisible"). We shall look first at the connectives individually.

A sentence which is modified by the word "not" is called the **negation** of the original sentence. For example, "2 is not a prime" is the negation of "2 is a prime," and "It is not the case that 2 is a prime and 6 is a composite number" is the negation of "2 is a prime and 6 is a composite number." It is because the latter sentence is composite that grammatical usage forces one to use the phrase "It is not the case that" instead of simply the word "not."

The word "and" is used to join two sentences to form a composite sentence which is called the **conjunction** of the two sentences. For example, the sentence "The sun is shining, and it is cold outside" is the conjunction of the sentences "The sun is shining" and "It is cold outside." In ordinary language various words, such as "but," are used as approximate synonyms for "and"; however, we shall ignore possible differences in shades of meaning which might accompany the use of one in place of the other.

A sentence formed by connecting two sentences with the word "or" is called the **disjunction** of the two sentences. We shall always assume that "or" is used in the inclusive sense (in legal documents this is often expressed by the barbarism "and/or"). Recall that we interpreted "or" in this way in the definition of the union of two sets.

From two sentences we may construct one of the form "If . . . , then . . ."; this is called a **conditional** sentence. The sentence immediately following "If" is the **antecedent,** and the sentence immediately following "then" is the **consequent.** For example, "If $2 > 3$, then $3 > 4$" is a conditional sentence with "$2 > 3$" as antecedent and "$3 > 4$" as consequent. Several other idioms in English which are generally regarded as having the same meaning as "If P, then Q" (where P and Q are sentences) are:

P implies Q;*
P only if Q;
P is a sufficient condition for Q;
Q, provided that P;
Q if P;
Q is a necessary condition for P.

The words "if and only if" are used to obtain from two sentences a **biconditional** sentence. We regard the biconditional

P if and only if Q

as having the same meaning as

If P, then Q, and if Q, then P;
Q is a necessary and sufficient condition for P.

By introducing letters "P," . . . to stand for prime sentences, a special symbol for each connective, and parentheses, as may be needed for punctuation, the connective structure of a composite sentence can be displayed in an effective manner. Our choice of symbols for the connectives is as follows:

$\sim$ for "not";
$\wedge$ for "and";
$\vee$ for "or";
$\rightarrow$ for "if . . . , then . . .";
$\leftrightarrow$ for "if and only if."

---

* In the hope of avoiding confusion, we shall restrict our use of "implies" to the technical sense defined in §2.4.

Thus, if P and Q are sentences, then

$$\sim P, \; P \wedge Q, \; P \vee Q, \; P \to Q, \; P \leftrightarrow Q$$

are, respectively, the negation of P, the conjunction of P and Q, and so on. Following are some concrete examples of analyzing the connective structure of composite sentences in terms of constituent prime sentences.

Examples

2.1.1. The sentence

> 2 is a prime, and 6 is a composite number

may be symbolized by

$$P \wedge C,$$

where P is "2 is a prime" and C is "6 is a composite number."

2.1.2. The sentence

> If either the Pirates or the Cubs lose and the Giants win, then the Dodgers will be out of first place and, moreover, I will lose a bet,

is a conditional, so it may be symbolized in the form

$$\ldots \to \ldots$$

The antecedent is composed from the three prime sentences P ("The Pirates lose"), C ("The Cubs lose"), and G ("The Giants win"), and the consequent is the conjunction of D ("The Dodgers will be out of first place") and B ("I will lose a bet"). The original sentence may be symbolized in terms of these prime sentences by

$$((P \vee C) \wedge G) \to (D \wedge B).$$

2.1.3. The sentence

> If either labor or management is stubborn, then the strike will be settled iff the government obtains an injunction, but troops are not sent into the mills.

is a conditional. The antecedent is the disjunction of L ("Labor is stubborn") and M ("Management is stubborn"). The consequent is a biconditional whose lefthand member is S ("The strike will be settled") and whose righthand member is the conjunction of G ("The government obtains an injunction") and the negation of R ("Troops are sent into the mills"). So, the original sentence may be symbolized by

$$(L \vee M) \to (S \leftrightarrow (G \wedge (\sim R))).$$

Exercises

2.1.1. Translate the following composite sentences into symbolic notation, using letters to stand for the prime components (which here we understand to mean sentences which contain no connectives).

- (a) Either it is raining or someone left the shower on.
- (b) If it is foggy tonight, then either John must stay home or he must take a taxi.
- (c) John will sit, and he or George will wait.
- (d) John will sit and wait, or George will wait.
- (e) I will go either by bus or by taxi.
- (f) Neither the North nor the South won the Civil War.
- (g) If, and only if, irrigation ditches are dug will the crops survive; should the crops not survive, then the farmers will go bankrupt and leave.
- (h) If I am either tired or hungry, then I cannot study.
- (i) If John gets up and goes to school, he will be happy; and if he does not get up, he will not be happy.

2.1.2. Let C be "Today is clear," R be "It is raining today," S be "It is snowing today," and Y be "Yesterday was cloudy." Translate into acceptable English the following:

(a) $C \to (\sim(R \land S))$;	(d) $(Y \to R) \lor C$;
(b) $Y \leftrightarrow C$;	(e) $C \leftrightarrow ((R \land (\sim S)) \lor Y)$;
(c) $Y \land (C \lor R)$;	(f) $(C \leftrightarrow R) \land ((\sim S) \lor Y)$.

§2.2. The Statement Calculus. Truth Tables

Earlier we agreed that by a statement we would understand a declarative sentence which is either true or false but not both. We call "truth" and "falsity" truth values and abbreviate them by T and F, respectively. Thus a statement has exactly one of T and F as a truth value. We consider next the assignment of truth values to composite sentences whose constituent prime sentences are statements.

On the basis of the usual meaning of "not," if a statement is true, its negation is false, and vice versa. For example, if S is the true statement (has truth value T) "The moon is a satellite of the earth," then $\sim S$ is false (has truth value F).

By convention, the conjunction of two statements is true when, and only when, both of its constituent statements are true. For example, "3 is a prime, and $2 + 2 = 5$" is false because "$2 + 2 = 5$" is a false statement.

Having agreed that the connective "or" would be understood in the inclusive sense, standard usage classifies a disjunction as false when, and only when, both constituent statements are false.

Truth-value assignments of the sort which we are making can be summarized concisely by **truth tables,** wherein are displayed the truth values of the composite statement for all possible assignments of truth values to the constituent statements. Table 2.1 displays the truth tables for the five sentential connectives.

Table 2.1.

Negation			Conjunction			Disjunction	
P	$\sim P$	P	Q	$P \wedge Q$	P	Q	$P \vee Q$
T	F	T	T	T	T	T	T
F	T	T	F	F	T	F	T
		F	T	F	F	T	T
		F	F	F	F	F	F

Conditional			Biconditional		
P	Q	$P \rightarrow Q$	P	Q	$P \leftrightarrow Q$
T	T	T	T	T	T
T	F	F	T	F	F
F	T	T	F	T	F
F	F	T	F	F	T

The motivation for the truth-value assignments made for the conditional in Table 2.1 is the fact that, as intuitively understood, $P \rightarrow Q$ is true iff Q is deducible from P in some way. So, if P is true and Q is false, we want $P \rightarrow Q$ to be false, which accounts for the second line of the table. Next, suppose that Q is true. Then, independently of P and its truth value, it is plausible to assert that $P \rightarrow Q$ is true. This reasoning suggests the assignments made in the first and third lines of the table. To justify the fourth line, consider the statement $(P \wedge Q) \rightarrow P$. We expect this to be true regardless of the choice of P and Q. But, if P and Q are both false, then $P \wedge Q$ is false, and we are led to the conclusion that if both antecedent and consequent are false, a conditional is true.

The table for the biconditional is determined by that for conjunction and the conditional, once it is agreed that $P \leftrightarrow Q$ means the same as $(P \rightarrow Q) \wedge (Q \rightarrow P)$.

These five truth tables are to be understood as definitions; they are the customary definitions adopted for mathematics. We have made merely a feeble attempt to make them seem plausible on the basis of meaning. It is an immediate consequence of these definitions that if P and Q are statements, then so are each of $\sim P$, $P \wedge Q$, $P \vee Q$, $P \to Q$, and $P \leftrightarrow Q$. It follows that any composite sentence whose prime components (that is, those prime sentences which appear in the composite sentence) are statements is itself a statement; if the truth values of the prime components are known, then the truth value of the composite statement can be determined in a mechanical way.

Examples

2.2.1. Suppose that a composite statement is symbolized by

$$(P \vee Q) \to (R \leftrightarrow (\sim S))$$

and that the truth values of P, Q, R, and S are T, F, F, and T, respectively. Then the value of $P \vee Q$ is T, that of $\sim S$ is F, that of $R \leftrightarrow (\sim S)$ is T, and, hence, that of the original statement is T, as a conditional having a true antecedent and a true consequent. Such a calculation can be made quickly by writing the truth value of each prime statement underneath it and the truth value of each composite constructed under the connective involved. Thus, for the above we would write out the following, where, for study purposes, we have put successive steps on successive lines.

$$
\begin{array}{ccccc}
(P & \vee & Q) \to (R \leftrightarrow (\sim & S)) \\
\text{T} & \text{F} & \text{F} & \text{T} \\
& \text{T} & & \text{F} \\
& & \text{T} & \\
& & \text{T} &
\end{array}
$$

2.2.2. Consider the following argument.

> If prices are high, then wages are high. Prices are high or there are price controls. Further, if there are price controls, then there is not an inflation. There is an inflation. Therefore, wages are high.

Suppose that we are in agreement with each of the first four statements (the premises). Must we accept the fifth statement (the conclusion)? To answer this, let us first symbolize the argument, using letters "P," "W," "C," and "I" in the obvious way. Thus, P is the sentence "Prices are high." Then we may present it as follows:

$$
\begin{array}{l}
P \to W \\
P \vee C \\
C \to (\sim I) \\
\underline{I} \\
W
\end{array}
$$

To assume that we are in agreement with the premises amounts to the assignment of the value T to the statements above the line. The question posed then can be phrased as: If the premises have value T, does the conclusion have value T? The answer is in the affirmative. Indeed, since I and $C \to (\sim I)$ have value T, the value of C is F according to the truth table for the conditional. Hence, P has value T (since $P \lor C$ is T) and, therefore, W has value T (since $P \to W$ is T).

2.2.3. We consider the conjunction

$$(P \lor C) \land (C \to (\sim I))$$

of two of the statements in Example 2.2.2. In general, the truth value which such a statement will receive is dependent on the assignments made to the prime statements involved. It is realistic to assume that, during periods of changing economic conditions, the appropriate truth value assignments to one or more of P, C, and I will change from T to F or vice versa. Thus the question may arise: What combinations of truth values of P, C, and I will give $(P \lor C) \land (C \to (\sim I))$ value T or value F? This can be answered by the examination of a truth table (Table 2.2) in which there appears the truth value of the composite statement for *every* possible assignment (2^3) of truth values to P, C, and I. This is called the truth table for the given statement. Each line includes an assignment of values to P, C, and I, along with the associated value of $(P \lor C) \land (C \to (\sim I))$. The latter may be computed as in the first example above. However, short cuts in filling out the complete table will certainly occur to the reader as he proceeds.

Table 2.2.

P	C	I	$(P \lor C) \land (\to(\sim I))$
T	T	T	F
T	T	F	T
T	F	T	T
T	F	F	T
F	T	T	F
F	T	F	T
F	F	T	F
F	F	F	F

2.2.4. If P is "2 is a prime" and L is "Logic is fun," there is nothing to prohibit our forming such composite statements as

$$P \lor L, \; P \to L, \; (\sim P) \to (P \lor L).$$

Since both P and L have truth values (clearly, both are T), these composite statements have truth values which we can specify. One's initial reaction to such nonsense might be that it should be prohibited—that the formation of conjunctions, conditionals, and so on, should be permitted only if the component statements are related in content or subject. However, it requires no lengthy reflection to realize the difficulties involved in characterizing such obscure notions. It is much simpler to take the easy way out, namely, to permit the formation of composite statements from any statements. On the basis of meaning, this amounts to nonsense sometimes, but no harm results. Our concern is with the formulation of principles of valid reasoning. In applications to systematic reasoning, composite statements which amount to gibberish simply will not occur.

Exercises

2.2.1. Suppose that the statements P, Q, R, and S are assigned the truth values T, F, F, and T, respectively. Find the truth value of each of the following statements.

(a) $(P \vee Q) \vee R$.

(b) $P \vee (Q \vee R)$.

(c) $R \rightarrow (S \wedge P)$.

(d) $P \rightarrow (R \rightarrow S)$.

(e) $P \rightarrow (R \vee S)$.

(f) $(P \vee R) \leftrightarrow (R \wedge (\sim S))$.

(g) $(S \leftrightarrow P) \rightarrow ((\sim P) \vee S)$.

(h) $(Q \wedge (\sim S)) \rightarrow (P \leftrightarrow S)$.

(i) $(R \wedge S) \rightarrow (P \rightarrow ((\sim Q) \vee S))$.

(j) $((P \vee (\sim Q)) \vee (R \rightarrow (S \wedge (\sim S)))$.

2.2.2. Construct the truth table for each of the following statements.

(a) $P \rightarrow (P \rightarrow Q)$.

(b) $(P \vee Q) \leftrightarrow (Q \vee P)$.

(c) $P \rightarrow \sim(Q \wedge R)$.

(d) $(P \rightarrow Q) \leftrightarrow ((\sim P) \vee Q)$.

(e) $(P \rightarrow (Q \wedge R)) \vee ((\sim P) \wedge Q)$.

(f) $(P \wedge Q) \rightarrow ((Q \wedge (\sim Q)) \rightarrow (R \wedge Q))$.

2.2.3. Suppose the value of $P \rightarrow Q$ is T; what can be said about the value of $((\sim P) \wedge Q) \leftrightarrow (P \vee Q)$?

2.2.4. (a) Suppose the value of $P \leftrightarrow Q$ is T; what can be said about the values of $P \leftrightarrow (\sim Q)$ and $(\sim P) \leftrightarrow Q$?

(b) Suppose the value of $P \leftrightarrow Q$ is F; what can be said about the values of $P \leftrightarrow (\sim Q)$ and $(\sim P) \leftrightarrow Q$?

2.2.5. For each of the following, determine whether the information given is sufficient to decide the truth value of the statement. If the information is enough, state the truth value. If it is insufficient, show that both truth values are possible.

(a) $(P \rightarrow Q) \rightarrow R$.
$\quad\quad\quad\quad\quad\mathsf{T}$

(b) $P \wedge (Q \rightarrow R)$.
$\quad\quad\quad\quad\mathsf{T}$

(c) $P \vee (Q \rightarrow R)$.
$\quad\quad\quad\quad\quad\mathsf{T}$

(d) $(\sim(P \vee Q)) \leftrightarrow ((\sim P) \wedge (\sim Q))$.
$\quad\quad\quad\mathsf{T}$

(e) $(P \rightarrow Q) \rightarrow ((\sim Q) \rightarrow \sim P))$.
$\quad\quad\quad\mathsf{T}$

(f) $(P \wedge Q) \rightarrow (P \vee S)$.
$\quad\quad\mathsf{T}\quad\quad\quad\quad\quad\mathsf{F}$

2.2.6. In Example 2.1.3 we symbolized the statement

> If either labor or management is stubborn, then the strike will be settled iff the government obtains an injunction, but troops are not sent into the mills,

as

$$(L \lor M) \to (S \leftrightarrow (G \land (\sim R))).$$

By a truth-value analysis, determine whether this statement is true or false under each of the following assumptions.

(a) Labor is stubborn, management is not, the strike will be settled, the government obtains an injunction, and troops are sent into the mills.

(b) Both labor and management are stubborn, the strike will not be settled, the government fails to obtain an injunction, and troops are sent into the mills.

2.2.7. Referring to the statement in the preceding exercise, suppose it is agreed that

> If the government obtains an injunction, then troops will be sent into the mills. If troops are sent into the mills, then the strike will not be settled. The strike will be settled. Management is stubborn.

Determine whether the statement in question is true or not.

§2.3.   The Statement Calculus. Validity

The foregoing is intended to suggest the nature of the statement calculus, namely, the analysis of those logical relations among sentences which depend solely on their composition from constituent sentences using sentential connectives. The setting for such an analysis includes the presence of an initial set of sentences (the "prime sentences") and the following two assumptions:

(i) Each prime sentence is a statement; that is, there may be assigned to a prime sentence a truth value.

(ii) Each sentence under consideration is composed from prime sentences using sentential connectives and, for a given assignment of truth values to these prime sentences, receives a truth value in accordance with the truth tables given earlier for negation, conjunction, and so on.

With this in mind let us make a fresh start on the statement calculus. By a **statement letter** we shall mean a capital Roman letter with or without a subscript from late in the alphabet: "P," "Q," . . . . Expressions built up from statement letters by appropriate applications of the sentential connectives will be called **statement forms.** Precisely,

(1) All statement letters are statement forms.

(2) If A and B are statement forms, then so are $(\sim A)$, $(A \vee B)$, $(A \wedge B)$, $(A \to B)$, and $(A \leftrightarrow B)$.

(3) An expression is a statement form only if it can be shown to be a statement form on the basis of clauses (1) and (2).*

A complicated statement form involves a large number of parentheses. To simplify the appearance of statement forms and thereby make them easier to read, we adopt the following conventions for the omission of parentheses. First, we shall omit the outermost pair of parentheses of a form. Second, when a form contains only one binary connective (that is, one of $\vee$, $\wedge$, $\to$, or $\leftrightarrow$), parentheses are omitted by association from the left. For example, $P \wedge Q \wedge R \wedge P$ stands for $((P \wedge Q) \wedge R) \wedge P$. Third, the connectives are ranked in order of decreasing scope as $\leftrightarrow$, $\to$, $\vee$, $\wedge$, $\sim$, and parentheses are eliminated according to the rule that, first, $\sim$ applies to the smallest form following it, then $\wedge$ is to connect the smallest statement forms on either side, and similarly for $\vee$, $\to$, and $\leftrightarrow$. In applying this rule to occurrences of the same connective, one proceeds from left to right. The following examples illustrate the conventions.

$$P \wedge Q \to R \text{ means } (P \wedge Q) \to R;$$
$$\sim P \wedge Q \text{ means } (\sim P) \wedge Q;$$
$$\sim P \to Q \leftrightarrow R \text{ means } ((\sim P) \to Q) \leftrightarrow R;$$
$$P \leftrightarrow Q \wedge R \wedge S \vee \sim R \to S \text{ means}$$
$$P \leftrightarrow ((((Q \wedge R) \wedge S) \vee (\sim R)) \to S).$$

Not every statement form can be represented without the use of parentheses. For example, the parentheses in $P \to (Q \to R)$ and in $\sim(P \wedge Q)$ cannot be eliminated using our conventions.

One may think of a statement letter as naming some prime sentence and of a statement form which is not a statement letter as a name of a composite sentence. By working with the formal analogues of prime and composite sentences, we eliminate all traces of meaning and internal structure of sentences. Thereby we can concentrate on the subject matter of the statement calculus as described in the first sentence of

---

* In more detail, B is a statement form iff there is a finite sequence $A_1, A_2, \ldots, A_n$ ($n \geq 1$) such that $B = A_n$ and, if A_i ($1 \leq i \leq n$) is either a statement letter or is a negation, disjunction, conjunction, conditional, or biconditional, constructed from previous expressions in the sequence. Note that we use capital Roman letters with or without subscripts from the initial part of the alphabet to stand for arbitrary statement forms.

this section and, in particular, describe the contribution of the statement calculus to a theory of inference. The first step in this direction is
the agreement that a statement letter is assigned one of T and F as truth
value. Secondly, the assignment of a truth value to a statement form
whose constituent statement letters have been assigned truth values is
made via the truth tables for the sentential connectives in Table 2.1.
In other words, the truth values of a statement form are defined by
induction in accordance with those tables.* If the statement letters in a
form A are $P_1, P_2, \ldots, P_n$, then the definition of the truth values of A
in terms of truth values of $P_1, P_2, \ldots, P_n$ can be exhibited in a truth
table like Table 2.2. There are 2^n rows in such a table, each row ex-
hibiting one possible assignment of T's and F's to $P_1, P_2, \ldots, P_n$.

The statement calculus is concerned with the truth values of statement forms in terms of truth-value assignments to the constituent statement letters and the interrelations of the truth values of statement forms
having some statement letters in common. As we proceed in this study,
it will appear that those forms whose truth value is T for every assignment of truth values to its statement letters occupy a central position.
A form whose value is T for all possible assignments of truth values to
its statement letters is a **tautology** or, alternatively, such a form is
(logically) valid (in the statement calculus). We shall often write

$$\models A$$

for "A is valid" or "A is a tautology." † Whether or not a form A is a
tautology can be determined by an examination of its truth table. If
the statement letters in A are $P_1, P_2, \ldots, P_n$, then A is a tautology iff
its value is T for each of the 2^n assignments of T's and F's to $P_1, P_2, \ldots, P_n$.
For example, $P \to P$ and $P \wedge (P \to Q) \to Q$ are tautologies, whereas
$P \to (Q \to R)$ is not, as examination of Tables 2.3, 2.4, and 2.5 will
show.

In a natural language like English or in a formal theory (see §3.5),
a sentence which arises from a tautology by substitution of sentences for

---

* An appeal to Theorem 1.12.2 need not be made for justifying such a definition by induction. Thinking of the natural numbers as the objects which can be generated one by one from
0 by the successor operation (the so-called constructive approach), one may view a definition
by induction as a recipe for finding, for any natural number n, the value of a function f.
First, $f(0)$ is given. Then for any natural number n, $f(n')$ is computed from $f(n)$ and possibly n
by the recipe. As for the definition which provoked this footnote, the induction is on the
number of connectives. Having completed the nth stage in the assignment of a truth value
to a form, one of the tables serves as a recipe for completing the next stage.

† This symbol $\models$ for validity was apparently first used by Kleene in this connection.

Table 2.3.

P	$P \to P$
T	T
F	T

Table 2.4.

P	Q	$P \wedge \cdot(P \to Q) \to Q$		
T	T	T	T	T
T	F	F	F	T
F	T	F	T	T
F	F	F	T	T

Table 2.5.

P	Q	R	$P \to (Q \to R)$	
T	T	T	T	T
T	T	F	F	F
T	F	T	T	T
T	F	F	T	T
F	T	T	T	T
F	T	F	T	F
F	F	T	T	T
F	F	F	T	T

all statement letters—occurrences of the same letter being replaced by the same sentence—is said to be **logically true** (according to the statement calculus). Such a sentence may be said to be true by virtue of its truth-functional construction alone. An example is the English sentence "If it is cloudy or cold and it is not cold, then it is cloudy," which results by substitution from the tautology $(P \vee Q) \wedge (\sim P) \to Q$.

The definition of validity provides us with a mechanical way to decide whether a given statement form is valid, namely, the computation and examination of its truth table. Although it may be tedious, this method can always be used to test a proposed form for validity. But, clearly, it is an impractical way to discover tautologies. This state of affairs has prompted the derivation of rules for generating tautologies from tautologies. The knowledge of a limited number of simple tautologies and several such rules make possible the derivation of a great variety of valid forms. We develop next several such rules and then implement them with a list of useful tautologies.

THEOREM 2.3.1. Let B be a statement form and let B^* be the form resulting from B by the substitution of a form A for all occurrences of a statement letter P occurring in B. If $\models B$, then $\models B^*$.

Proof. For an assignment of values to the statement letters of B^*, there results a value $v(A)$ of A and a value $v(B^*)$ of B^*. Now $v(B^*) = v(B)$, the value of B for a particular assignment of values to its statement letters, including the assignment of $v(A)$ to P. If B is valid, then $v(B)$ and hence $v(B^*)$ is always T. That is, if B is valid, then so is B^*.

Examples

2.3.1. From Table 2.6 it follows that $\vDash P \vee Q \leftrightarrow Q \vee P$. Hence, by Theorem 2.3.1, $\vDash (R \rightarrow S) \vee Q \leftrightarrow Q \vee (R \rightarrow S)$. A direct verification of this result (Table 2.7) using the reasoning employed in the proof of Theorem 2.3.1 should clarify matters, if need be. To explain the relationship of Table 2.7 to Table 2.6,

<table>
<tr><th colspan="2">Table 2.6.</th><th colspan="4">Table 2.7.</th></tr>
</table>

P	Q	$P \vee Q \leftrightarrow Q \vee P$	R	S	Q	$(R \rightarrow S) \vee Q \leftrightarrow Q \vee (R \rightarrow S)$
T	T	T				
T	F	T				
F	T	F T T T T T F	T	F	T	F T T F
F	F	T				

we discuss the displayed line of Table 2.7. There was entered first (at two places) the value F of $R \rightarrow S$ for the assignment of T to R and F to S. Then the value T assigned to Q was entered twice. The rest of the computation is then a repetition of that appearing in the third line of Table 2.6 after the entries underlined there have been made.

2.3.2. The practical importance of Theorem 2.3.1 is that it provides a method to establish the validity of a form without dissecting it all the way down to its statement letters. An illustration will suffice to describe what we have in mind. Suppose the question arises whether the statement form

$$(R \vee S) \wedge ((R \vee S) \rightarrow (P \wedge Q)) \rightarrow (P \wedge Q)$$

is a tautology. The answer is in the affirmative, with Theorem 2.3.1 supplying the justification, as soon as it is recognized that the form in question has "the same form" as the tautology $P \wedge (P \rightarrow Q) \rightarrow Q$ (Table 2.4), in the sense that it results from this tautology upon substituting $R \vee S$ for P and $P \wedge Q$ for Q.

We introduce next a binary relation for statement forms based on a comparison of their truth tables. For the definition to be applicable to every pair of forms, we note that if $P_1, P_2, \ldots, P_n$ are the statement letters occurring in a statement form A, then A may be considered to be composed from an extended list $P_1, P_2, \ldots, P_n, P_{n+1}, \ldots, P_m$ of letters and a truth table for A constructed in terms of this extended list. Now, if A and B are statement forms, let $P_1, P_2, \ldots, P_m$ be a list of the statement letters appearing in either A or B. We define A to be **equivalent** to B, symbolized

$$A \text{ eq } B,^*$$

* We emphasize that this notation is part of English.

iff the truth tables of A and B, constructed in terms of $P_1, P_2, \ldots, P_m$ are the same. As illustrations, we infer from Tables 2.8 and 2.9 that $(P \rightarrow Q)$ eq $\sim P \vee Q$ and P eq $P \wedge (Q \vee \sim Q)$.

Table 2.8.

P	Q	$P \rightarrow Q$	$\sim P \vee Q$
T	T	T	T
T	F	F	F
F	T	T	T
F	F	T	T

Table 2.9.

P	Q	P	$P \wedge (Q \vee \sim Q)$
T	T	T	T
T	F	T	T
F	T	F	F
F	F	F	F

It is left as an exercise to prove that eq is an equivalence relation on every set of statement forms and, further, that it has the following substitutivity property: If C_A is a statement form containing a specific occurrence of the form A, and C_B is the result of replacing this occurrence of A by a form B, then

$$\text{if } B \text{ eq } A, \text{ then } C_B \text{ eq } C_A.$$

Henceforth, equivalent forms will be regarded as interchangeable within the context of notions definable in terms of truth tables only, and the substitution property will often be employed without comment. Equivalence of statement forms can be characterized in terms of the concept of a valid form in the following way.

THEOREM 2.3.2. $\vDash A \leftrightarrow B$ iff A eq B.

Proof. Let $P_1, P_2, \ldots, P_m$ be the totality of statement letters appearing in A and B. For a given assignment of truth values to these, the first part of the computation of the value of $A \leftrightarrow B$ consists of computing the values of A and B, after which the computation is concluded by applying the table for the biconditional. According to this table, the value of $A \leftrightarrow B$ is T iff the values computed for A and B are the same.

COROLLARY. Let C_A be a formula containing a specified occurrence of the formula A and let C_B be the result of replacing this occurrence of A by a formula B. If $\vDash A \leftrightarrow B$, then $\vDash C_A \leftrightarrow C_B$. If $\vDash A \leftrightarrow B$ and $\vDash C_A$, then $\vDash C_B$.

This proof is left as an exercise.

THEOREM 2.3.3. If $\models A$ and $\models A \rightarrow B$, then $\models B$.

Proof. Let $P_1, P_2, \ldots, P_m$ be the totality of statement letters appearing in A and B. For a given assignment of truth values to these, the first part of the computation of the value of $A \rightarrow B$ consists of computing the values of A and B, after which the computation is completed by applying the table for the conditional. The assumptions $\models A$ and $\models A \rightarrow B$ imply that both the value obtained for A and that for $A \rightarrow B$ are $\top$. According to the table for $A \rightarrow B$, this implies that B must also have the value $\top$. Since this is the case for all assignments of values to $P_1, P_2, \ldots, P_m$, B is valid.

As the next theorem we list a collection of tautologies. It is not intended that these be memorized, but rather they should be used for reference. That many of the biconditionals listed are tautologies should be highly plausible on the basis of meaning, together with Theorem 2.3.1. That each is a tautology may be demonstrated by constructing a truth table for it, regarding the letters present as statement letters. Then, once it is shown that the value is $\top$ for all assignments of values to the letters, an appeal is made to the substitution rule of Theorem 2.3.1 to remove the restriction that the letters be statement letters. In the exercises for this section, the reader is asked to establish the validity of some of the later statement forms by applying one or more of Theorems 2.3.1, 2.3.2, and 2.3.3 to tautologies appearing earlier in the list.

THEOREM 2.3.4.

Tautological Conditionals

1. $\models A \wedge (A \rightarrow B) \rightarrow B$.
2. $\models {\sim}B \wedge (A \rightarrow B) \rightarrow {\sim}A$.
3. $\models {\sim}A \wedge (A \vee B) \rightarrow B$.
4. $\models A \rightarrow (B \rightarrow A \wedge B)$.
5. $\models A \wedge B \rightarrow A$.
6. $\models A \rightarrow A \vee B$.
7. $\models (A \rightarrow B) \wedge (B \rightarrow C) \rightarrow (A \rightarrow C)$.
8. $\models (A \wedge B \rightarrow C) \rightarrow (A \rightarrow (B \rightarrow C))$.
9. $\models (A \rightarrow (B \rightarrow C)) \rightarrow (A \wedge B \rightarrow C)$.
10. $\models (A \rightarrow B \wedge {\sim}B) \rightarrow {\sim}A$.
11. $\models (A \rightarrow B) \rightarrow (A \vee C \rightarrow B \vee C)$.
12. $\models (A \rightarrow B) \rightarrow (A \wedge C \rightarrow B \wedge C)$.
13. $\models (A \rightarrow B) \rightarrow ((B \rightarrow C) \rightarrow (A \rightarrow C))$.
14. $\models (A \leftrightarrow B) \wedge (B \leftrightarrow C) \rightarrow (A \leftrightarrow C)$.

Tautological Biconditionals

15. $\models A \leftrightarrow A$.
16. $\models \sim\sim A \leftrightarrow A$.
17. $\models (A \leftrightarrow B) \leftrightarrow (B \leftrightarrow A)$.
18. $\models (A \rightarrow B) \wedge (C \rightarrow B) \leftrightarrow (A \vee C \rightarrow B)$.
19. $\models (A \rightarrow B) \wedge (A \rightarrow C) \leftrightarrow (A \rightarrow B \wedge C)$.
20. $\models (A \rightarrow B) \leftrightarrow (\sim B \rightarrow \sim A)$.
21. $\models A \vee B \leftrightarrow B \vee A$.   21'. $\models A \wedge B \leftrightarrow B \wedge A$.
22. $\models (A \vee B) \vee C \leftrightarrow$ 22'. $\models (A \wedge B) \wedge C \leftrightarrow$
$\qquad\qquad A \vee (B \vee C)$. $\qquad\qquad A \wedge (B \wedge C)$.
23. $\models A \vee (B \wedge C) \leftrightarrow$ 23'. $\models A \wedge (B \vee C) \leftrightarrow$
$\qquad (A \vee B) \wedge (A \vee C)$. $\qquad (A \wedge B) \vee (A \wedge C)$.
24. $\models A \vee A \leftrightarrow A$.   24'. $\models A \wedge A \leftrightarrow A$.
25. $\models \sim(A \vee B) \leftrightarrow$ 25'. $\models \sim(A \wedge B) \leftrightarrow$
$\qquad\qquad \sim A \wedge \sim B$. $\qquad\qquad \sim A \vee \sim B$.

Tautologies for Elimination of Connectives

26. $\models A \rightarrow B \leftrightarrow \sim A \vee B$.
27. $\models A \rightarrow B \leftrightarrow \sim(A \wedge \sim B)$.
28. $\models A \vee B \leftrightarrow \sim A \rightarrow B$.
29. $\models A \vee B \leftrightarrow \sim(\sim A \wedge \sim B)$.
30. $\models A \wedge B \leftrightarrow \sim(A \rightarrow \sim B)$.
31. $\models A \wedge B \leftrightarrow \sim(\sim A \vee \sim B)$.
32. $\models (A \leftrightarrow B) \leftrightarrow (A \rightarrow B) \wedge (B \rightarrow A)$.

Example

2.3.3. We present an alternative approach to the statement calculus in terms
of functions. The point of departure is the observation that if P_1, P_2, ..., P_n
are the statement letters occurring in a statement form A, then A provides a
rule for associating with each ordered n-tuple of T's and F's one of T and F, upon
assigning to P_i the ith coordinate of the n-tuple ($1 \leq i \leq n$). If we set $V = \{T,F\}$
and define a **truth function** as a function on V^n into V, our observation can be
rephrased as: A expresses a truth function. The truth table for A displays the
truth function which A expresses. Truth functions will be designated by such
symbols as

$$f(p_1, p_2, \ldots, p_n), \qquad g(q_1, q_2, \ldots, q_n), \text{ and so on.}$$

Note that we depart from our practice of designating functions by single letters
and use notation heretofore reserved for function values.* Our excuse is that

---

* This departure can be avoided by using lambda notation for functions.

composition of functions can be described more simply. For example, the notation

$$f(p_1, \ldots, p_{i-1}, g(q_1, \ldots, q_m), p_{i+1}, \ldots, p_n)$$

is self-explanatory as a function obtained by composition from the truth function f of n arguments and g of m arguments. We shall refer to this function as that obtained by substitution of g for the ith variable in f. Clearly, such combinations of truth functions are again truth functions.

An alternative approach to the statement calculus can be given in terms of truth functions: There are 2^{2^n} different truth functions of n variables. Of the four for $n = 1$, that one whose value at T is F and whose value at F is T we denote by $\sim p$. Among the sixteen truth functions of two arguments occur the four presented in Table 2.10. The reason for the denotations chosen should be clear.

Table 2.10.

	$\wedge (p,q)$	$\vee (p,q)$	$\rightarrow (p,q)$	$\leftrightarrow (p,q)$
$\langle \mathsf{T},\mathsf{T}\rangle$	T	T	T	T
$\langle \mathsf{T},\mathsf{F}\rangle$	F	T	F	F
$\langle \mathsf{F},\mathsf{T}\rangle$	F	T	T	F
$\langle \mathsf{F},\mathsf{F}\rangle$	F	F	T	T

It is of interest that all truth functions of any number of arguments can be generated from those functions named so far with the operation of function composition in the form described above. Indeed, the three functions $\sim p$, $\vee (p,q)$, and $\wedge (p,q)$ suffice. To prove this let $f(p_1, p_2, \ldots, p_n)$ be a truth function. If the value of f is F for all values of $p_1, p_2, \ldots, p_n$ (that is, f is the constant function F), then it is equal to

$$\wedge (\ldots \wedge (\wedge (\wedge (p_1, \sim p_1), \wedge (p_2, \sim p_2)), \wedge (p_3, \sim p_3)) \ldots, \wedge (p_n, \sim p_n)).$$

Otherwise, f assumes the value T at least once. For each member of the domain of f such that f is assigned the value T, form the function

$$\wedge (\ldots \wedge (\wedge (q_1, q_2), q_3) \ldots, q_n),$$

where q_i is p_i or $\sim p_i$ according as p_i has the value T or F. Then we contend that f is equal to the function obtained by composing all such functions using the "disjunction function" $\vee (p,q)$. For example, if $f(p,q)$ takes the value F when $p = q = \mathsf{T}$ and the value T otherwise, then

$$f(p,q) = \vee (\vee (\wedge (p, \sim q), \wedge (\sim p, q)), \wedge (\sim p, \sim q)).$$

The reader can verify this and supply a proof of the general statement.

Actually, each of the pairs $\sim p$, $\wedge (p,q)$ and $\sim p$, $\vee (p,q)$ is adequate to generate all truth functions using function composition, since (with equality being that of functions)

$$\vee(p,q) = \sim(\wedge(\sim p,\sim q)) \text{ and } \wedge(p,q) = \sim(\vee(\sim p,\sim q)).$$

The same is true of the pair $\sim p$, $\rightarrow(p,q)$, as the reader can verify. Although no member of any of the three pairs can be discarded to obtain a single function which generates all truth functions, such functions exist. For example, the "stroke function" $|(p,q)$ of two arguments whose value is $\top$ except at $\langle\top,\top\rangle$ where its value is F suffices. To prove this, it is sufficient to show, for example, that both $\sim p$ and $\vee(p,q)$ can be expressed in terms of it.

The proof sketched above that the three truth functions $\sim p$, $\vee(p,q)$, and $\wedge(p,q)$ generate all truth functions can easily be converted into a proof of the following statement: Every truth function is expressed by a statement form (in the sense explained at the beginning of this example) involving the connectives $\sim$, $\vee$, and $\wedge$. Via this route one concludes, for example, that a statement form which expresses the function $f(p,q)$ defined above is

$$(P \wedge \sim Q) \vee (\sim P \wedge Q) \vee (\sim P \wedge \sim Q).$$

We conclude this section with the description of a powerful method for obtaining tautologies from scratch. Initially we consider only statement forms composed from statement letters $P_1, P_2, \ldots, P_n$ using $\sim$, $\wedge$, and $\vee$. The **denial**, A_d, of such a form A is the form resulting from A by replacing each occurrence of $\wedge$ by $\vee$ and vice versa, and replacing each occurrence of P_i by an occurrence of $\sim P_i$ and vice versa. As illustrations of denials in the present context we note that the denial of $P \vee Q$ is $\sim P \wedge \sim Q$ and the denial of $\sim(\sim P \wedge Q)$ is $\sim(P \vee \sim Q)$. The theorem relating denials and tautologies follows.

THEOREM 2.3.5. Let A be a statement form composed from statement letters using only $\sim$, $\wedge$, and $\vee$. Let A_d be the denial of A. Then $\models \sim A \leftrightarrow A_d$.

A proof of this assertion can be given by induction on the number of symbols appearing in a form. We forego this, but include in the first example below a derivation of an instance of the theorem. Another example describes the extension of the theorem to the case of a form which involves $\rightarrow$ or $\leftrightarrow$.

Examples

2.3.4. An instance of Theorem 2.3.5 is the assertion that

$$\models \sim((\sim P \vee Q) \vee (Q \wedge (R \vee \sim P))) \leftrightarrow$$
$$(P \wedge \sim Q) \wedge (\sim Q \vee (\sim R \wedge P)),$$

or, in other words, that the lefthand side and the righthand side of the biconditional are equivalent formulas. Using the properties of transitivity and substitutivity of equivalence, this is established below. Each step is justified by the indicated part of Theorem 2.3.4 (in view of Theorem 2.3.2).

$$\sim((\sim P \vee Q) \vee (Q \wedge (R \vee \sim P)))$$
$$\text{eq } \sim(\sim P \vee Q) \wedge \sim(Q \wedge (R \vee \sim P)) \qquad (25)$$
$$\text{eq } (\sim\sim P \wedge \sim Q) \wedge (\sim Q \vee \sim(R \vee \sim P)) \qquad (25, 25')$$
$$\text{eq } (\sim\sim P \wedge \sim Q) \wedge (\sim Q \vee (\sim R \wedge \sim\sim P)) \qquad (25)$$
$$\text{eq } (P \wedge \sim Q) \wedge (\sim Q \vee (\sim R \wedge P)) \qquad (16)$$

2.3.5. Using tautology 32 in Theorem 2.3.4, we can derive from a statement form in which $\leftrightarrow$ appears an equivalent form in which $\leftrightarrow$ is absent. For instance,

$$P \leftrightarrow (Q \wedge R) \text{ eq } (P \rightarrow Q \wedge R) \wedge (Q \wedge R \rightarrow P).$$

That is, $\leftrightarrow$ can be eliminated from any form. Similarly, using tautology 26 or 27, $\rightarrow$ can be eliminated from any form. Thus, any form A is equivalent to a form B composed from statement letters using $\sim$, $\wedge$, and $\vee$. Then we may define the denial of A to be the denial of B.

2.3.6. According to the preceding example, $\leftrightarrow$ and $\rightarrow$ can be eliminated from any form. Using tautology 29 it is possible to eliminate $\vee$ or (with tautology 31), equally well, $\wedge$. That is, any statement form is equivalent to one composed from statement letters using $\sim$ and $\vee$ or using $\sim$ and $\wedge$. This conclusion was obtained in Example 2.3.3.

2.3.7. From tautology 22 follows the general associative law for $\vee$, which asserts that however parentheses are inserted in $A_1 \vee A_2 \vee \cdots \vee A_n$ to render it unambiguous, the resulting statement forms are equivalent. From tautology 22' follows the corresponding result for $\wedge$.

Exercises

2.3.1. Suppose that $P_1, P_2, \ldots, P_n$ are the statement letters occurring in A. Show that the truth table of A, regarded as having $P_1, \ldots, P_n, \ldots, P_m$ as statement letters, can be divided into 2^{m-n} parts, each a duplicate of the truth table for A computed with $P_1, P_2, \ldots, P_n$ as its statement letters.

2.3.2. Prove that eq is an equivalence relation on every set of statement forms and that it has the substitutivity property stated in the text.

2.3.3. Prove the result stated in the Corollary to Theorem 2.3.2.

2.3.4. Prove that a statement form containing only the connective $\leftrightarrow$ is a tautology iff each statement letter occurs an even number of times.

2.3.5. Derive each of tautologies 28–31 from earlier tautologies in Theorem 2.3.4, using properties of equivalence for formulas. As an illustration, we derive tautology 27 from earlier ones. From 26, $A \rightarrow \sim B$ eq $\sim A \vee \sim B$, and, in turn,

$\sim A \lor \sim B$ eq $\sim(A \land B)$ by 25'. Hence, $A \to \sim\sim B$ eq $\sim(A \land \sim B)$. Using 16 it follows that $A \to B$ eq $\sim(A \land \sim B)$, which amounts to 27.

Note: Example 2.3.3 is the basis for Exercises 2.3.6 through 2.3.12.

2.3.6. Generate each of the following truth functions from $\sim p$, $\land(p,q)$, and $\lor(p,q)$.

(a) p	q	r	$f(p,q,r)$		(b) p	q	r	$g(p,q,r)$
T	T	T	T		T	T	T	T
T	T	F	F		T	T	F	F
T	F	T	T		T	F	T	F
T	F	F	F		T	F	F	F
F	T	T	F		F	T	T	T
F	T	F	F		F	T	F	F
F	F	T	T		F	F	T	F
F	F	F	T		F	F	F	T

2.3.7. Supply the missing details in the proof that every truth function can be generated from $\sim p$, $\lor(p,q)$, and $\land(p,q)$ by composition.

2.3.8. Convert the proof of the assertion in Exercise 2.3.7 to a proof that every truth function is the truth function determined by a statement form involving the connectives $\sim$, $\lor$, and $\land$.

2.3.9. A statement form is in **disjunctive normal form** if it is a disjunction consisting of one or more disjuncts, each of which is a conjunction of one or more statement letters and negations of statement letters. Some examples are

$(P \land Q) \lor (\sim P \lor R), \quad (P \land Q \land \sim Q) \lor (R \land \sim P) \lor R, \quad \sim P, \quad P \land Q.$

A form is in **conjunctive normal form** if it is a conjunction of one or more conjuncts, each of which is a disjunction of one or more statement letters and negations of statement letters. The result asserted in Exercise 2.3.8 may be restated as: Every statement form A is logically equivalent to one in disjunctive normal form. By applying this result to $\sim A$, prove that A is also equivalent to a form in conjunctive normal form.

2.3.10. Find logically equivalent disjunctive and conjunctive normal forms for $(P \to Q) \lor (\sim P \land R)$ and $((P \to \sim Q) \land R) \lor (\sim P \leftrightarrow R)$. Hint: Instead of employing the method used in proving that this is always possible, use the techniques described in Example 2.3.5 to eliminate $\leftrightarrow$ and $\to$ from any statement form and then use other tautologies from Theorem 2.3.4.

2.3.11. Prove that every truth function is determined by a statement form containing as connectives only $\sim$ and $\land$, or only $\sim$ and $\lor$, or only $\sim$ and $\to$.

2.3.12. Prove that neither the pair $\to$, $\lor$ nor the pair $\sim$, $\leftrightarrow$ is adequate to produce all truth functions.

2.3.13. Instead of completing a truth table to determine the values of a statement form, an arithmetic procedure may be used. The basis for this approach is the representation of the basic statement forms by arithmetic functions as designated in Table 2.11.* When the value T (respectively, F) is assigned to a

Table 2.11.

Form	Arithmetic representation
$\sim P$	$1 + P$
$P \wedge Q$	PQ
$P \vee Q$	$P + Q + PQ$
$P \rightarrow Q$	$1 + P(1 + Q)$
$P \leftrightarrow Q$	$1 + P + Q$

statement letter in a form—for example, P—the value 1 (respectively, 0) is assigned to the variable P in the associated arithmetical representation. Further, values of the arithmetical function are computed as in ordinary arithmetic, with one exception: namely, $1 + 1 = 0$.

In each case a calculation shows that when the form takes the value T (respectively, F), then its arithmetical representation takes the value 1 (respectively, 0). In these terms, tautologies are represented by functions which are identically 1. For example, that $\models P \vee \sim P$ follows from observation that $P \vee \sim P$ is represented by $P + (1 + P) + P(1 + P)$ which is equal to (the constant function) 1, since in the algebra at hand $x + x = 0$, $x(1 + x) = 0$, and $x^2 = x$ for all x. To prove that tautology 1 of Theorem 2.3.4 (regarding A and B as statement letters) actually is a tautology, we form first [corresponding to $A \wedge (A \rightarrow B)$],

$$A(1 + A(1 + B))$$

which is equal to AB. Then to the entire form in tautology 1 corresponds the function $1 + AB(1 + B)$ which is equal to 1.

Prove some of the tautologies in Theorem 2.3.4 by this method.

2.3.14. (a) With Exercise 2.3.13 in mind, show that the function $1 + PQ$ is an arithmetical representation of the truth function $P|Q$ defined in Example 2.3.3.

(b) The result in (a) may be reformulated as follows: Every mapping on $\{0,1\}^n$ into $\{0,1\}$ can be generated from the mapping $f: \{0,1\}^2 \rightarrow \{0,1\}$ such that $f(x,y) = 1 + xy$. Show that the same is true of $g: \{0,1\}^3 \rightarrow \{0,1\}$ where $g(x,y,z) = 1 + x + y + xyz$.

---

* The origin of this correspondence is to be found in the relationship between Boolean algebras and Boolean rings. The reader who wishes to investigate this should first acquaint himself with §4.6 and then study Birkhoff (1967), pp. 47–49.

§2.4. The Statement Calculus. Consequence

In the introduction to this chapter, we said that it was a function of logic to provide principles of reasoning, that is, a theory of inference. In practical terms this amounts to supplying criteria for deciding in a mechanical way whether a chain of reasoning will be accepted as correct on the basis of its form. A chain of reasoning is simply a finite sequence of statements which are supplied to support the contention that the last statement in the sequence (the conclusion) may be inferred from certain initial statements (the premises). In everyday circumstances the premises of an inference are judged to be true (on the basis of experience, experiment, or belief). Acceptance of the premises of an inference as true and of the principles employed in a chain of reasoning from such premises as correct commits one to regard the conclusion at hand as true. In a mathematical theory the situation is different. There, one is concerned solely with the conclusions (the so-called "theorems" of the theory) which can be inferred from an assigned initial set of statements (the so-called "axioms" of the theory) according to rules which are specified by some system of logic. In particular, the notion of truth plays no part whatsoever in the theory proper. The contribution of the statement calculus to a theory of inference is just this: It provides a criterion, along with practical working forms thereof, for deciding when the concluding sentence (a statement) of an argument is to be assigned the value T if each premise of the argument is assigned the truth value T. This criterion is in the form of a definition. The statement form B is (**logical**) **consequence** of statement forms $A_1, A_2, \ldots, A_m$ (by the statement calculus), symbolized

$$A_1, A_2, \ldots, A_m \models B,$$

iff for every truth-value assignment to each of the statement letters $P_1, P_2, \ldots, P_n$ occurring in one or more of $A_1, A_2, \ldots, A_m$, and B, the form B receives the value T whenever every A receives the value T. In terms of truth tables, "$A_1, A_2, \ldots, A_m \models B$" means simply that if truth tables are constructed for $A_1, A_2, \ldots, A_m$, and B, from the list $P_1, P_2, \ldots, P_n$ of statement letters occurring in one or more of these statement forms, then B receives the value T at least for each assignment to the P's which make all A's simultaneously T.

Example

2.4.1. From an inspection of Table 2.12 we obtain the following three illustrations of our definition:

$$P, R, Q \wedge P \to \sim R \vDash \sim Q;$$ (line 3)
$$P, P \to R, R \vDash P \vee Q \to R;$$ (lines 1 and 3)
$$Q \wedge P \to \sim R, \sim Q, P \to R \vDash \sim(P \wedge Q).$$ (lines 3, 7, 8)

Table 2.12.

P	Q	R	$Q \wedge P \to \sim R$	$\sim Q$	$P \to R$	$P \vee Q \to R$	$\sim(P \wedge Q)$
T	T	T	F	F	T	T	F
T	T	F	T	F	F	F	F
T	F	T	T	T	T	T	T
T	F	F	T	T	F	F	T
F	T	T	T	F	T	T	T
F	T	F	T	F	T	F	T
F	F	T	T	T	T	T	T
F	F	F	T	T	T	T	T

A special case of the above definition deserves our attention. A statement form A is said to **(logically) imply** the form B if B is a consequence of A. According to the first part of the next theorem, A implies B iff the conditional $A \to B$ is valid.

THEOREM 2.4.1.

(I) $A \vDash B$ iff $\vDash A \to B$.

(II) $A_1, A_2, \ldots, A_m \vDash B$ iff $A_1 \wedge A_2 \wedge \cdots \wedge A_m \vDash B$ or, iff $\vDash A_1 \wedge A_2 \wedge \cdots \wedge A_m \to B(m \geq 2)$.

Proof. For (I), let $A \vDash B$. By the table for $\to$, $A \to B$ receives the value F iff A receives the value T, and, simultaneously, B receives the value F. From the hypothesis, this combination of values does not occur. Hence $A \to B$ always receives the value T, that is, $\vDash A \to B$. For the converse, let $\vDash A \to B$, and consider an assignment of values to the statement letters such that A receives the value T. Since $A \to B$ receives the value T, it follows from the table for $\to$ that B takes the value T, whence, $A \vDash B$.

The first assertion in (II) follows from the table for $\wedge$, and the second follows from the first by an application of (I).

COROLLARY. $A_1, \ldots, A_{m-1}, A_m \vDash B$ iff $A_1, \ldots, A_{m-1} \vDash A_m \to B$. More generally, $A_1, \ldots, A_{m-1}, A_m \vDash B$ iff $\vDash A_1 \to (A_2 \to (\cdots (A_m \to B) \cdots))$.

Proof. For $m = 1$, the first assertion is (I) of the theorem. So, assume that $A_1, \ldots, A_{m-1}, A_m \models B$ for $m > 1$. Then $\models (A_1 \wedge \cdots \wedge A_{m-1}) \wedge A_m \rightarrow B$, according to the theorem. From tautology 8 of Theorem 2.3.4 and Theorem 2.3.3, we deduce that $\models (A_1 \wedge \cdots \wedge A_{m-1}) \rightarrow (A_m \rightarrow B)$. According to (I) of the theorem, it follows that $A_1 \wedge \cdots \wedge A_{m-1} \models A_m \rightarrow B$ and hence, by (II), that $A_1, \ldots, A_{m-1} \models A_m \rightarrow B$. The converse is established by reversing the foregoing steps.

Finally, the second assertion follows by repeated application of the first.

Thus, the problem of what statements are consequences of others (by the statement calculus) is reduced to the problem of what statements are valid (which accounts for the importance of tautologies). On the other hand, there are reasons for approaching the concept of consequence directly. One reason is the possibility of converting the definition into a working form which resembles that used in mathematics to infer theorems from a set of axioms. Indeed, we can substantiate a working form as a sequence of statement forms (the last form being the desired consequence of the premises) such that the presence of each is justified by a rule, called a **rule of inference** (for the statement calculus). The basis for the rules of inference which we shall introduce is the following theorem.

THEOREM 2.4.2.

(I) $A_1, A_2, \ldots, A_m \models A_i$ for $i = 1, 2, \ldots, m$.

(II) If $A_1, A_2, \ldots, A_m \models B_j$ for $j = 1, 2, \ldots, p$, and if $B_1, B_2, \ldots, B_p \models C$, then $A_1, A_2, \ldots, A_m \models C$.

Proof. Part (I) is an immediate consequence of the definition of "$A_1, A_2, \ldots, A_m \models B$." For (II) we construct a truth table from the list $P_1, P_2, \ldots, P_n$ of all statement letters appearing in at least one of the A's, the B's, and C. Consider any row in which $A_1, A_2, \ldots, A_m$ each receive the value T. Then, by the hypotheses, each B has the value T, and hence C has the value T. That is, for each assignment of values to the P's such that every A takes the value T, formula C receives the value T. This is the desired conclusion.

With this result, a demonstration that a form B (the conclusion) is a consequence of statement forms $A_1, A_2, \ldots, A_m$ (the premises) may be presented as a **string** (that is, a finite sequence) of statement forms, the last of which is B and such that the presence of each form E is justified by an application of one of the following rules.

Rule p: The form E is a premise.

Rule t: There are statement forms $A, \ldots, C$ preceding E in the string such that $\vDash A \wedge \cdots \wedge C \rightarrow E$.

That is, we contend that $A_1, A_2, \ldots, A_m \vDash B$ if we can concoct a string

$$E_1, E_2, \ldots, E_r(= B)$$

of statement forms such that either each E is a premise (rule p) or there are preceding forms in the string such that if C is their conjunction, then $\vDash C \rightarrow E$ (rule t). Indeed, assuming that each entry in the displayed sequence can be so justified, we shall prove that

$$A_1, A_2, \ldots, A_m \vDash \text{(any } E \text{ in the sequence)}.$$

This is true of E_1 by Theorem 2.4.2 (I). Assume that each of $E_1, E_2, \ldots, E_{k-1}$ is a consequence of the A's; we prove that the same is true of the next form E_k. If E_k is a premise, then Theorem 2.4.2 (I) applies. Otherwise, there are statement forms preceding E_k such that if C is their conjunction, then $\vDash C \rightarrow E_k$. Let us say

$$\vDash E_{i_1} \wedge E_{i_2} \wedge \cdots \wedge E_{i_s} \rightarrow E_k.$$

Then, by Theorem 2.4.1(II),

$$E_{i1}, E_{i2}, \ldots, E_{i_s} \vDash E_k,$$

and, by assumption,

$$A_1, A_2, \ldots, A_m \vDash E_{ij}, \qquad j = 1, 2, \ldots, s.$$

Hence, by Theorem 2.4.2(II),

$$A_1, A_2, \ldots, A_m \vDash E_k.$$

We note, finally, that by an application of rule t any tautology may be entered in a derivation. Indeed, if $\vDash D$, then for any form A we have $\vDash A \rightarrow D$. Thus, D may be included in a derivation by an application of rule t wherein we take any premise as the "A."

The examples which follow illustrate the foregoing method for demonstrating that some form is a consequence of given statement forms. To make the method entirely definite, let us agree that when applying rule t, only the tautological conditionals which appear explicitly in Theorem 2.3.4 or are implicit in the biconditionals of that theorem (for example, $\vDash A \leftrightarrow A$ yields the tautological conditional $\vDash A \rightarrow A$ and

$\models \sim\sim A \leftrightarrow A$ yields $\models \sim\sim A \rightarrow A$ and $\models A \rightarrow \sim\sim A$) may be used. Admittedly, this is an arbitrary rule. Our excuse for making it is that it serves to make the game to be played a definite one.

Examples

2.4.2. We demonstrate that

$$A \vee B, A \rightarrow C, B \rightarrow D \models C \vee D.$$

An explanation of the numerals on the left is given below.

$\{1\}$	(1) $A \rightarrow C$	Rule p
$\{1\}$	(2) $A \vee B \rightarrow C \vee B$	Rule t; $\models$ (1) $\rightarrow$ (2) by tautology 11.
$\{3\}$	(3) $B \rightarrow D$	Rule p
$\{3\}$	(4) $C \vee B \rightarrow C \vee D$	Rule t; $\models$ (3) $\rightarrow$ (4) by tautology 11.
$\{1,3\}$	(5) $A \vee B \rightarrow C \vee D$	Rule t; $\models$ (2) $\wedge$ (4) $\rightarrow$ (5) by tautology 7.
$\{6\}$	(6) $A \vee B$	Rule p
$\{1,3,6\}$	(7) $C \vee D$	Rule t; $\models$ (6) $\wedge$ (5) $\rightarrow$ (7) by tautology 1.

The number in parentheses adjacent to each form designates both that form and the line of the derivation in which it appears. The set of numbers in braces for each line corresponds to the premises on which the form in that line depends. That is, the form in line n is a consequence of the premises designated by the numbers in braces in that line. Thus, the form in line 5 is a consequence of the premise in line 1 and the premise in line 3, and the form in line 7 is a consequence of the premises in lines 1, 3, and 6, that is, of all the premises. In particular, for a line which displays a premise there appears in braces at the left just the number of that line, since such a form depends on no other line. Using the brace notation in connection with the numerals on the left is deliberate in that it suggests that the form in that line is a consequence of the set of premises designated by those numbers.

We now rewrite the above derivation, incorporating some practical abbreviations. In this form the reader is called on to supply the tautologies employed.

$\{1\}$	(1) $A \rightarrow C$	p
$\{1\}$	(2) $A \vee B \rightarrow C \vee B$	1 t
$\{3\}$	(3) $B \rightarrow D$	p
$\{3\}$	(4) $C \vee B \rightarrow C \vee D$	3 t
$\{1,3\}$	(5) $A \vee B \rightarrow C \vee D$	2, 4 t
$\{6\}$	(6) $A \vee B$	p
$\{1,3,6\}$	(7) $C \vee D$	5, 6 t

2.4.3. As a more elaborate illustration we prove that

$$W \vee P \rightarrow I, I \rightarrow C \vee S, S \rightarrow U, \sim C \wedge \sim U \vDash \sim W$$

by the following string of thirteen statement forms.

{1}	(1) $\sim C \wedge \sim U$	p
{1}	(2) $\sim U$	1 t
{3}	(3) $S \rightarrow U$	p
{1,3}	(4) $\sim S$	2, 3 t
{1}	(5) $\sim C$	1 t
{1,3}	(6) $\sim C \wedge \sim S$	4, 5 t
{1,3}	(7) $\sim (C \vee S)$	6 t
{8}	(8) $W \vee P \rightarrow I$	p
{9}	(9) $I \rightarrow C \vee S$	p
{8,9}	(10) $W \vee P \rightarrow C \vee S$	8, 9 t
{1,3,8,9}	(11) $\sim (W \vee P)$	7, 10 t
{1,3,8,9}	(12) $\sim W \wedge \sim P$	11 t
{1,3,8,9}	(13) $\sim W$	12 t

We note that the foregoing takes the place of a truth table having $2^6 = 64$ lines for the purpose of verifying that

$$\vDash (W \vee P \rightarrow I) \wedge (I \rightarrow C \vee S) \wedge (S \rightarrow U) \wedge (\sim C \wedge \sim U) \rightarrow \sim W.$$

2.4.4. Many theorems in mathematics have the form of a conditional, the assumptions being the axioms of the theory under development. The symbolic form of such a theorem is

$$A_1, A_2, \ldots, A_m \vDash B \rightarrow C,$$

where the A's are the axioms and $B \rightarrow C$ is the consequence asserted. In order to prove such a theorem, it is standard practice to adopt B as a further assumption and then infer that C is a consequence. Thereby it it is implied that

$$A_1, A_2, \ldots, A_m \vDash B \rightarrow C \text{ iff } A_1, A_2, \ldots, A_m, B \vDash C.$$

This is correct according to the Corollary to Theorem 2.4.1. It is convenient to formulate this as a third rule of inference, the rule of conditional proof, for the statement calculus.

Rule cp: The form $B \rightarrow C$ is justified in a derivation having $A_1, A_2, \ldots, A_m$ as premises if it has been established that C is a consequence of $A_1, A_2, \ldots, A_m$, and B.

As an illustration of the use of this rule we prove that

$$A \rightarrow (B \rightarrow C), \sim D \vee A, B \vDash D \rightarrow C.$$

{1}	(1) $A \rightarrow (B \rightarrow C)$	p
{2}	(2) $\sim D \vee A$	p
{3}	(3) B	p

{4}	(4) D	p (introducing "D" as an additional premise)
{2,4}	(5) A	2, 4 t
{1,2,4}	(6) $B \to C$	1, 5 t
{1,2,3,4}	(7) C	3, 6 t
{1,2,3}	(8) $D \to C$	4, 7 cp

The usefulness of the braced numbers to show precisely what premises enter into the derivation of the form in that line is clear.

2.4.5. Even if an alleged consequence of a set of premises does not have the form of a conditional, the application of the strategy as described in the preceding example may simplify a derivation. As an illustration we rework the first example, starting with the observation that the conclusion $C \lor D$ is equivalent to $\sim C \to D$. This suggests adding $\sim C$ as a premise and hoping that D can be derived as a consequence of this and the other premises. An advantage gained thereby is the addition of a simple assumption. The derivation follows.

{1}	(1) $A \lor B$	p
{2}	(2) $A \to C$	p
{3}	(3) $B \to D$	p
{4}	(4) $\sim C$	p
{2,4}	(5) $\sim A$	2, 4 t
{1,2,4}	(6) B	1, 5 t
{1,2,3,4}	(7) D	3, 6 t
{1,2,3}	(8) $\sim C \to D$	4, 7 cp
{1,2,3}	(9) $C \lor D$	8 t

2.4.6. Each of the tautological conditionals in Theorem 2.3.4 generates a rule of inference, namely, the instance of rule t, which is justified by reference to that tautology alone. For example, tautology 1 in Theorem 2.3.4 determines the rule

From A and $A \to B$ to infer B.

This is called the **rule of detachment** or **modus ponens.** In a textbook devoted to logic, names for many rules of inference of this sort will be found. Probably modus ponens is the one used most frequently in derivations.

The foregoing method for demonstrating that a statement form B is a consequence (as specified by the statement calculus) of forms $A_1, A_2, \ldots, A_m$ is of interest only if the set $\{A_1, A_2, \ldots, A_m\}$ is (simultaneously) **satisfiable,** by which is meant that there exists an assignment of truth values to the constituent statement letters such that the A_i's simultaneously receive truth value T. For if a set $\{A_1, A_2, \ldots, A_m\}$ of statement forms is **nonsatisfiable,** that is, for every assignment of truth values to the constituent statement letters at least one A_i receives value F, then *every* B is a consequence, according to the definition of this notion. Upon defining a statement form to be a **contradiction** (or **logically false**) iff it always takes the value F (as, for instance, does $A \wedge \sim A$), we can give the following characterization of nonsatisfiability.

THEOREM 2.4.3. A set $\{A_1, A_2, \ldots, A_m\}$ of statement forms is nonsatisfiable iff a contradiction can be derived as a consequence of the set.

Proof. If $\{A_1, A_2, \ldots, A_m\}$ is nonsatisfiable, then every form, and so in particular a contradiction, is a consequence. Conversely, assume that $A_1, A_2, \ldots, A_m \vDash C$, where C is a contradiction. Then $\vDash A_1 \wedge A_2 \wedge \cdots \wedge A_m \to C$ and the conclusion follows from the truth tables for the conditional and conjunction.

Contradictions also play an important role in the method of **indirect proof** (also called **proof by contradiction** or **reductio ad absurdum proof**). The basis for this type of proof is the following result.

THEOREM 2.4.4. $A_1, A_2, \ldots, A_m \vDash B$ if a contradiction can be derived as a consequence of $\{A_1, A_2, \ldots, A_m, \sim B\}$.

Proof. Assume that $A_1, A_2, \ldots, A_m, \sim B \vDash C$, for some contradiction C. Then, by Theorem 2.4.3, $\{A_1, A_2, \ldots, A_m, \sim B\}$ is nonsatisfiable. Hence, for each assignment of truth values to the constituent statement letters such that each A_i receives value T, $\sim B$ receives value F and so B receives value T. Hence, $A_1, A_2, \ldots, A_m \vDash B$.

Exercises

2.4.1. By an examination of Table 2.12, justify the conclusions drawn in Example 2.4.1.

2.4.2. Complete each of the following demonstrations of consequence by supplying the tautologies employed and the numbering scheme discussed in Example 2.4.2.

(a) $A \to B,\ \sim(B \vee C) \vDash \sim A$

$A \to B$
$\sim(B \vee C)$
$\sim B \wedge \sim C$
$\sim B$
$\sim A$

(c) $(A \wedge B) \vee (C \wedge D)$,
$\qquad\qquad\qquad A \to \sim A \vDash C$

$A \to \sim A$
$\sim A \vee \sim A$
$\sim A$
$\sim A \vee \sim B$
$\sim(A \wedge B)$
$(A \wedge B) \vee (C \wedge D)$
$C \wedge D$
C

(b) $A \rightarrow B, C \rightarrow B, D \rightarrow$
$$A \lor C, D \models B$$

$D \rightarrow A \lor C$
D
$A \lor C$
$A \rightarrow B$
$C \rightarrow B$
$A \lor C \rightarrow B$
B

(d) $A \rightarrow (C \rightarrow B), \sim D \lor A,$
$$C \models D \rightarrow B$$

$\sim D \lor A$
D
A
$A \rightarrow (C \rightarrow B)$
$C \rightarrow B$
C
B
$D \rightarrow B$

2.4.3. Justify each of the following, using only rules p and t.
(a) $\sim A \lor B, C \rightarrow \sim B \models A \rightarrow \sim C$.
(b) $A \rightarrow (B \rightarrow C), C \land D \rightarrow E, \sim F \rightarrow (D \land \sim E) \models A \rightarrow (B \rightarrow F)$.
(c) $A \lor B \rightarrow C \land D, D \lor E \rightarrow F \models A \rightarrow F$.
(d) $A \rightarrow (B \land C), \sim B \lor D, (E \rightarrow \sim F) \rightarrow \sim D,$
$$B \rightarrow (A \land \sim E) \models B \rightarrow E.$$
(e) $(A \rightarrow B) \land (C \rightarrow D), (B \rightarrow E) \land (D \rightarrow F),$
$$\sim (E \land F), A \rightarrow C \models \sim A.$$

2.4.4. Try to shorten your proofs of Exercise 2.4.3(a), (b), (c), (d) using rule cp (along with rules p and t).

2.4.5. Can the rule of conditional proof be used to advantage in Exercise 2.4.3(e)? Justify your answer.

§2.5. The Statement Calculus. Applications

We now turn to some household applications of the theory of inference which we have discussed. Thus we revert, for the most part, to the terminology and point of view evidenced in §§2.1 and 2.2. Although capital letters will usually denote meaningful statements, it is only their truth values that concern us and so the concepts, notation, and results of §§2.3 and 2.4 may be used.

Usually the circumstances accompanying the presentation of an argument include the audience having the privilege of accepting or rejecting the contention that some statement B is a consequence of statements $A_1, A_2, \ldots, A_m$. In this event, the man who thinks for himself will want to prove either that B is a consequence of the A's or that the argument is **invalid,** that is, that there can be made an assignment of truth values to the statement letters at hand such that simultaneously each A receives value T, and B receives value F. The most expedient way to cope with the entire matter is this: Assume that B has value F and that each A has value T, and analyze the consequences insofar as necessary assignments of truth values to statement letters are concerned. Such an analysis will lead to either a contradiction, which proves that B is a consequence of the A's, or an assignment to each statement letter such that all assumptions are satisfied, which proves that the argument is invalid.

The foregoing technique may be employed to decide whether a given statement form is a consequence of some set of statement forms. Since it proceeds so quickly, it tends to undercut that developed in §2.4. The earlier method has its place, however, for it forces a practitioner to think about what follows from what (in the statement calculus) solely on the basis of form. Thinking in these terms requires repeated references to the tautologies listed in Theorem 2.3.4. This is good, for instances of these are commonplace in proofs in mathematics, and the reader should learn to recognize them as such. As an illustration, tautology 20 justifies the familiar conclusion that if the contrapositive, $\sim Q \to \sim P$, of $P \to Q$ is a consequence of A, then so is $P \to Q$.

Examples

2.5.1. Consider the following argument.

> If I go to my first class tomorrow, then I must get up early, and if I go to the dance tonight, I will stay up late. If I stay up late and get up early, then I will be forced to exist on only five hours of sleep. I simply cannot exist on only five hours of sleep. Therefore, I must either miss my first class tomorrow or not go to the dance.

To investigate the validity of this argument, we symbolize it using letters for prime statements. Let C be "I (will) go to my first class tomorrow," G be "I must get up early," D be "I (will) go to the dance tonight," S be "I will stay up late," and E be "I can exist on five hours of sleep." Then the premises may be symbolized as

$$(C \to G) \land (D \to S),$$
$$S \land G \to E,$$
$$\sim E,$$

and the desired conclusion as

$$\sim C \lor \sim D.$$

Following the method of analysis suggested above, we assume that $\sim C \lor \sim D$ has value F and that each premise has value T. Then each of C and D must have value T. Further, according to the first premise, both G and S have value T. This and the second premise imply that E has value T. But this contradicts the assumption that the third premise has value T. Thus we have proved that $\sim C \lor \sim D$ is a consequence of the premises.

2.5.2. Suppose it is asserted that

$$A \to B, C \to D, A \lor C \models B \land D.$$

Assume that $B \land D$ has value F and each premise has value T. The first assumption is satisfied if T is assigned to B and F is assigned to D. Then C has value F, and A has value T. With these assignments, each premise receives value T, and $B \land D$ takes value F. Hence the argument is invalid.

2.5.3. We illustrate the usefulness of Theorem 2.4.3 in proving the nonsatisfiability of a set of statements. Such a proof follows the same pattern as one devised to establish the correctness of an argument in all but one respect: in a proof of the correctness of an argument the final line, which is the conclusion, is assigned in advance, whereas, in a proof of nonsatisfiability the final line is any contradiction. For example, suppose that it is a question of the satisfiability of a set of statements which may be symbolized as follows:

$$A \leftrightarrow B, B \to C, \sim C \lor D, \sim A \to D, \sim D.$$

We adopt these as a set of premises and investigate what inferences can be made.

{1}	(1) $A \leftrightarrow B$	p
{2}	(2) $B \to C$	p
{3}	(3) $\sim C \lor D$	p
{4}	(4) $\sim A \to D$	p
{5}	(5) $\sim D$	p
{4,5}	(6) $\sim\sim A$	4, 5 t
{4,5}	(7) A	6 t
{1,2}	(8) $A \to C$	1, 2 t
{1,2,4,5}	(9) C	7, 8 t
{3,5}	(10) $\sim C$	3, 5 t
{1,2,3,4,5}	(11) $C \land \sim C$	9, 10 t

We conclude that the set is nonsatisfiable.

2.5.4. We could introduce a further rule of inference based on Theorem 2.4.4. Alternatively, we may employ the rule of conditional proof and the tautology $\models (\sim B \to C \land \sim C) \to B$ to justify an indirect proof. As an illustration, we rework Example 2.5.1, starting with the negation of the desired conclusion as an additional premise.

(1) $(C \to G) \land (D \to S)$	p
(2) $S \land G \to E$	p
(3) $\sim E$	p
(4) $\sim(\sim C \lor \sim D)$	p
(5) $C \land D$	
(6) C	
(7) $C \to G$	
(8) G	

(9) $D \rightarrow S$
(10) D
(11) S
(12) $S \wedge G$
(13) E
(14) $E \wedge \sim E$
(15) $\sim(\sim C \vee \sim D) \rightarrow E \wedge \sim E$
(16) $(\sim(\sim C \vee \sim D) \rightarrow (E \wedge \sim E)) \rightarrow \sim C \vee \sim D$
(17) $\sim C \vee \sim D$

It is left as an exercise to supply the missing details.

Exercises

Use the method discussed in this section to prove the validity or invalidity, whichever the case might be, of the arguments in Exercises 2.5.1 through 2.5.12. For those which are valid, construct a formal proof. In every case use the letters suggested for symbolizing the argument.

2.5.1. Either I shall go home or stay and have a drink. I shall not go home. Therefore I shall stay and have a drink. (H,S)

2.5.2. If John stays up late tonight, he will be dull tomorrow. If he doesn't stay up late tonight, then he will feel that life is not worth living. Therefore, either John will be dull tomorrow or he will feel that life is not worth living. (S,D,L)

2.5.3. Wages will increase only if there is inflation. If there is inflation, then the cost of living will increase. Wages will increase. Therefore, the cost of living will increase. (W,I,C)

2.5.4. If 2 is a prime, then it is the least prime. If 2 is the least prime, then 1 is not a prime. The number 1 is not a prime. Therefore, 2 is a prime. (P,L,N)

2.5.5. Either John is exhausted or he is sick. If he is exhausted, then he is contrary. He is not contrary. Therefore, he is sick. (E,S,C)

2.5.6. If it is cold tomorrow, I'll wear my heavy coat if the sleeve is mended. It will be cold tomorrow, and that sleeve will not be mended. Therefore, I'll not wear my heavy coat. (C,H,S)

2.5.7. If the races are fixed or the gambling houses are crooked, then the tourist trade will decline, and the town will suffer. If the tourist trade decreases, then the police force will be happy. The police force is never happy. Therefore, the races are not fixed. (R,H,D,S,P)

2.5.8. If the Dodgers win, then Los Angeles will celebrate, and if the White Sox win, Chicago will celebrate. Either the Dodgers will win or the White Sox will win. However, if the Dodgers win, then Chicago will not celebrate, and if the White Sox win, Los Angeles will not celebrate. So, Chicago will celebrate if and only if Los Angeles does not celebrate. (D,L,W,C)

2.5.9. Either Sally and Bob are the same age or Sally is older than Bob. If Sally and Bob are the same age, then Nancy and Bob are not the same age. If Sally is older than Bob, then Bob is older than Walter. Therefore, either Nancy and Bob are not the same age or Bob is older than Walter. (S,O,N,W)

2.5.10. If 6 is a composite number, then 12 is a composite number. If 12 is a composite number, then there exists a prime greater than 12. If there exists a prime greater than 12, then there exists a composite number greater than 12. If 2 divides 6, then 6 is a composite number. The number 12 is composite. Therefore, 6 is a composite number. (S,W,P,C,D)

2.5.11. If I take the bus, then if the bus is late, I'll miss my appointment. If I miss my appointment and start to feel downcast, then I should not go home. If I don't get that job, then I'll start to feel downcast and should go home. Therefore, if I take the bus, then if the bus is late, I will get that job. (B,L,M,D,H,J)

2.5.12. If Smith wins the nomination, he will be happy, and if he is happy, he is not a good compaigner. But if he loses the nomination, he will lose the confidence of the party. He is not a good compaigner if he loses the confidence of the party. If he is not a good compaigner, then he should resign from the party. Either Smith wins the nomination or he loses it. Therefore, he should resign from the party. (N,H,C,P,R)

2.5.13. Investigate the following sets of premises for satisfiability. If you conclude that a set is nonsatisfiable by assigning truth values, then reaffirm this using Theorem 2.4.3 and vice versa. Substantiate each assertion of the satisfiability of a set of premises by suitable truth-value assignments.

(a) $A \rightarrow \sim(B \wedge C)$
$D \vee E \rightarrow G$
$G \rightarrow \sim(H \vee I)$
$\sim C \wedge E \wedge H$

(b) $A \vee B \rightarrow C \wedge D$
$D \vee E \rightarrow G$
$A \vee \sim G$

(c) $(A \rightarrow B) \wedge (C \rightarrow D)$
$(B \rightarrow D) \wedge (\sim C \rightarrow A)$
$(E \rightarrow G) \wedge (G \rightarrow \sim D)$
$\sim E \rightarrow E$

(d) $(A \rightarrow B \wedge C) \wedge (D \rightarrow B \wedge E)$
$(G \rightarrow \sim A) \wedge H \rightarrow I$
$(H \rightarrow I) \rightarrow G \wedge D$
$\sim(\sim C \rightarrow E)$

(e) The contract is fulfilled if and only if the house is completed in February. If the house is completed in February, then we can move in March 1. If we can't move in March 1, then we must pay rent for March. If the contract is not fulfilled, then we must pay rent for March. We will not pay rent for March. (C,H,M,R)

2.5.14. Give an indirect proof of each of the following arguments.

(a) The second of Examples B in Section 2.4.

(b) The third of Examples B in Section 2.4.

(c) The first of Examples A in this section.

(d) Exercise 7 above.
(e) Exercise 11 above.
(f) Exercise 12 above.

2.5.15. Prove that if A, $\sim B \models C$ (a contradiction), then $A \models B$.

§2.6. First-Order Logic. Symbolizing Everyday Language

The theory of inference supplied by the statement calculus is inadequate for mathematics and, indeed, for everyday arguments. In support of this assertion consider the following inferences.

1. Every rational number is a real number. 3 is a rational number. Hence, 3 is a real number.
2. Some of Aristotle's followers like all of Aquinas' followers. None of Aristotle's followers like any idealist. Hence, none of Aquinas' followers are idealists.
3. All dogs are animals. Hence, the head of a dog is the head of an animal.

Certainly each conclusion is justified. The correctness of the inference in each case rests not only upon the truth-functional relations among the sentences involved, but also upon the internal structure of these sentences, as well as upon the meaning of such words as "every," "some," and "all." Thus what is needed is a way to dissect a sentence into sufficiently "fine" constituents so as to expose its internal structure. This is possible with the addition of the logical notions called *terms, predicates,* and *quantifiers.* These, together with the sentential connectives, provide adequate tools for symbolizing much of everyday and mathematical language in such a way as to make possible an analysis of an argument or a proof. We shall describe these three notions.

That part of the statement "Saul likes Paul" which is expressed by "likes Paul" is an example of a predicate. Instead of "likes Paul" we prefer to write

4. x likes Paul

where the letter "x" reserves a place for the missing subject. If in 4 we substitute for x the name of an appropriate object, there results a statement. Thus the predicate is a function whose values are statements (assuming a suitable domain for the function); 4, which we call a *predicate form*, is the recipe for computing values of this function. In logic the notion of a predicate is extended to include functions obtained in a similar way from statements by deleting the object (rather than the subject) or both subject and object. Thus we accept

5.   Saul likes x

and

6. x likes y

as (predicate forms) expressing predicates. The predicate expressed by **4** and that expressed by **5** are 1-place predicates. The predicate expressed by **6** is a 2-place predicate. These are illustrations of our definition of an **n-place predicate** as a function of n variables whose values are statements. We include 0 as a permissible value, understanding by a 0-place predicate a statement.

In §1.2, 1-place predicates were called properties; with a property is associated a set via the principle of abstraction. Similarly, 2-place predicates are often called binary relations—if $A(x,y)$ expresses some 2-place predicate [examples of such a form include **6** above, "$x < y$," and "x divides y"], then one can associate with the predicate those ordered pairs $\langle a,b \rangle$ such that $A(a,b)$ is a true statement. A corresponding remark may be made about n-place predicates.

When a predicate contains more than one occurrence of the same letter, it is understood that the same name is to be filled in for each occurrence. For example, from the 3-place predicate form

7. x likes y and y likes z,

we can get the statement "Paul likes Saul and Saul likes Mary" but not "Paul likes Saul and Jane likes Adam." But it is permissable to replace distinct letters with the same name of an individual. For example, from **7** we can get the statement "Paul likes Saul and Saul likes Paul."

We turn next to a description of the objects that will be called terms. We shall find use for letters and symbols as names of specific, well-defined objects; that is, we shall use letters and symbols for proper names. Letters and symbols used for this purpose are called individual constants. Individual constants form one class of terms. As our first illustration, we note that "3" is an individual constant, being a name of the numeral 3. Again, "Winston Churchill" is an individual constant. In order to achieve a compact notation we shall use a letter from the beginning of the alphabet to stand for a proper name if there is no accepted symbol for it. For example, we might let

$$a = \text{Winston Churchill}$$

if we intend to translate the sentence

Winston Churchill was a great statesman

into symbolic form.

Proper names are often rendered by a "description" which we take to be a name which by its own structure unequivocally identifies the object of which it is a name. For example,

The first president of the United States

and

The real number x such that for all real numbers y, $xy = y$

are descriptions. If we let

$$b = \text{George Washington,}$$

then we may write

$$b = \text{The first president of the United States.}$$

Further, we have

$$1 = \text{the real number } x \text{ such that for all } y, xy = y.$$

In the discussion of predicates above, the letters "x," "y," and "z" were used as place markers for individual constants. It is standard practice in mathematics to introduce letters for this purpose. For example, in order to determine those real numbers such that the square of the number minus the number is equal to twelve, one will form the equation $x^2 - x = 12$, thereby regarding "x" as a place holder for the name of any such (initially unknown) number. Again, as it is normally understood, the "x" in such an equation as

$$\sin^2 x + \cos^2 x = 1$$

reserves a place for the name of any real or, indeed, complex number. Although one is accustomed to refer to "x" in "$x^2 - x = 12$" as an unknown and as a variable in "$\sin^2 x + \cos^2 x = 1$," there is the common underlying feature that in each context "x" is a placeholder. In logic, letters such as "x" and "y," possibly with subscripts, used in this role are called "individual variables." The grammatical function of variables, which constitute another class of terms, is similar to that of pronouns and common nouns in everyday language.

We proceed by defining a **term** as an expression which either names or describes some object, or results in a name or description of some object when the individual variables occurring in it are replaced by names or descriptions. The earlier classification of individual constants and variables as terms anticipated this definition. Functions provide for a further class. For example, if h denotes the function that assigns to each animal its head, then $h(y)$ is a term. Again, using the familiar symbol for the binary operation of addition, each of

$$3 + 5, \ x + 5, \ x + y$$

is a term.

We interrupt our exposition with some examples of partial translations of everyday language into symbolic form that can be given in terms of the notions presented thus far.

Examples

2.6.1. The sentence

8.   Every rational number is a real number

may be translated as

9.   For every x, if x is a rational number, then x is a real number.

In ordinary grammar, "is a real number" is the predicate of **8**. In the translation **9** the added predicate expressed by "x is a rational number" replaces the common noun "rational number." Using "$Q(x)$" for "x is a rational number" and "$R(x)$" for "x is a real number," we may symbolize **9** as

10.   For every x, $Q(x) \rightarrow R(x)$.

Further, the statement "3 is a rational number" may be symbolized by

11. $Q(3)$.

In terms of symbolism available at the moment, **10** and **11** are the translations of the premises of the argument **1** at the beginning of this section.

2.6.2. The sentence

                 Some real numbers are rational

we translate as

      For some x, x is a real number and x is a rational number.

Using the abbreviations introduced above, this may be symbolized as

12.   For some x, $R(x) \land Q(x)$.

2.6.3. The sentence

13.   For some x, $R(x) \rightarrow Q(x)$

should have the same meaning as

14.   For some x, $\sim R(x) \vee Q(x)$,

since we have merely replaced "$R(x) \rightarrow Q(x)$" by its equivalent "$\sim R(x) \vee Q(x)$." Now **14** may be translated into words as

> There is something which is either not a real number or is a
> rational number.

Certainly, this statement [which has the same meaning as **13**] does not have the same meaning as **12**. Indeed, as soon as we exhibit an object which is not a real number we must subscribe to **13**. In summary, **13** and **12** have different meanings.

By assumption, on suitable assignments of values to the variables in a predicate form a statement results. For example, the form "x is a sophomore," yields the statement "John is a sophomore." A statement may also be obtained from the same form by prefixing it with the phrase "for every x":

15.   For every x, x is a sophomore.

No doubt, one would choose to rephrase this as

16.   Everyone is a sophomore.

The phrase "for every x" is called a **universal quantifier.** We regard "for every x," "for all x," and "for each x" as having the same meaning and symbolize each by

$$(\forall x) \text{ or } (x).$$

Using this symbol and "$S(x)$" for "x is a sophomore," we may symbolize **15** or **16** as

$$(x)S(x).$$

Similarly, prefixing $S(x)$ with the phrase "there exists an x (such that)" yields a statement which has the same meaning as "There are sophomores." The phrase "there exists an x" is called an **existential quantifier**. We regard "there exists an x," "for some x," and "for at least one x" as having the same meaning, and symbolize each by

$$(\exists x).$$

Thus, "$(\exists x)S(x)$" is the symbolic form of "There are sophomores."

Using the symbol introduced for the universal quantifier, we can now render "Every rational number is a real number" in its final form:

17. $$(x)(Q(x) \rightarrow R(x)).$$

Possibly it has already occurred to the reader that this means simply that $\mathbb{Q} \subseteq \mathbb{R}$. Indeed, if one recalls the definition of the inclusion relation for sets, it becomes clear that **17** is an instance of that definition. Further, we note that **17** is characteristic of statements of the form "Every so and so is a such and such."

Similarly, the sentence "Some real numbers are rational" may be translated as

$$(\exists x)(R(x) \wedge Q(x)).$$

The meaning of this sentence is simply that $\mathbb{R} \cap \mathbb{Q}$ is nonempty; that is, it is a symmetrical form of the original sentence. A mistake commonly made by beginners is to infer, since a statement of the form "Every so and so is a such and such" can be symbolized as in **17**, that the statement "Some so and so is a such and such" can be symbolized by

$$(\exists x)(R(x) \rightarrow Q(x)).$$

However, as is pointed out in Example 2.6.3, this should have the same meaning as

$$(\exists x)(\sim R(x) \vee Q(x)).$$

This should be accepted as true as soon as we exhibit an object which is not a real number. In particular, therefore, it has no relation to what it is intended to say, namely, that some real numbers are rational.

Examples

2.6.4. If $A(x)$ denotes a predicate in x, consider the following four statements.

(a) $(x)A(x)$; (c) $(x)(\sim A(x))$;
(b) $(\exists x)A(x)$; (d) $(\exists x)(\sim A(x))$.

We might translate these into words as follows:

(a) Everything has property A.
(b) Something has property A.
(c) Nothing has property A.
(d) Something does not have property A.

Now (d) is the denial of (a), and (c) is the denial of (b), on the basis of everyday meaning. Thus, for example, the existential quantifier may be defined in terms

of the universal quantifier by agreeing that "$(\exists x)A(x)$" is an abbreviation for
"$\sim((x)(\sim A(x)))$." *

2.6.5. Traditional logic emphasized four basic types of statements involving
quantifiers. Illustrations of these along with translations appear below. Two of
these translations have been discussed.

All rationals are reals.	$(x)(Q(x) \rightarrow R(x))$.
No rationals are reals.	$(x)(Q(x) \rightarrow \sim R(x))$.
Some rationals are reals.	$(\exists x)(Q(x) \wedge R(x))$.
Some rationals are not reals.	$(\exists x)(Q(x) \wedge \sim R(x))$.

2.6.6. If the symbols for negation and a quantifier modify a predicate form,
the order in which they appear is relevant. For example, the translation of

$$\sim(x)(x \text{ is mortal})$$

is "Not everyone is mortal" or "Someone is immortal," whereas the translation of

$$(x)(\sim(x \text{ is mortal}))$$

is "Everyone is immortal."

2.6.7. By prefixing a predicate form in several variables with a quantifier
(of either kind) for each variable, a statement results. For example, if it is
understood that all variables are restricted to the set of real numbers, then

$$(x)(y)(z)((x + y) + z = x + (y + z))$$

is the statement to the effect that addition is an associative operation. Again,

$$(x)(\exists y)(x^2 - y = y^2 - x)$$

translates into "For every (real number) x there is a (real number) y such that
$x^2 - y = y^2 - x$." This is a true statement. Notice, however, that

$$(\exists y)(x)(x^2 - y = y^2 - x),$$

obtained from the foregoing by interchanging the quantifiers, is a different,
indeed, a false statement.

2.6.8. We supplement the first remark in the preceding example with the
observation that a predicate form in several variables can also be reduced to a
statement by substituting values for all occurrences of some variables and applying quantifiers which pertain to the remaining variables. For example, the
(false) statement

$$(x)(x < 3)$$

results from the 2-place predicate "$x < y$" by substituting a value for y and
quantifying x.

---

* This abbreviation becomes more transparent if the range of x is a finite set. For instance,
suppose that $\{a,b,c\}$ is the range of x. Then $(x)A(x)$ means $A(a) \wedge A(b) \wedge A(c)$, whereas
$(\exists x)A(x)$ means $A(a) \vee A(b) \vee A(c)$, which is equivalent to (using 29 of Theorem 2.3.4)
$\sim(\sim A(a) \wedge \sim A(b) \wedge \sim A(c))$ which, in turn, may be rewritten as $\sim(x)(\sim A(x))$.

We conclude this section with the remark that there are no mechanical rules for translating sentences from English into symbolic form using the logical notation which has been introduced. In every case one must first decide on the meaning of the English sentence and then attempt to convey that same meaning in terms of predicates, quantifiers, and, possibly, individual constants.

As illustrations we shall symbolize arguments **2** and **3** at the beginning of this section. Abbreviate "x is Aristotle's follower" by "$A(x)$," "x is Aquinas' follower" by "$Q(x)$," "x is an idealist" by "$I(x)$," and "x likes y" by "$L(x,y)$." A symbolized version of **2** is then

$$(\exists x)(A(x) \wedge (y)(Q(y) \rightarrow L(x,y))), \qquad \text{(Premise)}$$
$$(x)(A(x) \rightarrow (y)(I(y) \rightarrow \sim L(x,y))), \qquad \text{(Premise)}$$
$$(x)(Q(x) \rightarrow \sim I(x)). \qquad \text{(Conclusion)}$$

For **3** abbreviate "x is a dog" by "$D(x)$," "x is an animal" by "$A(x)$," and "the head of x" by "$h(x)$." Then (3) may be symbolized as

$$(x)(D(x) \rightarrow A(x)), \qquad\qquad\qquad\qquad\qquad\qquad \text{(Premise)}$$
$$(x)((\exists y)(x = h(y) \wedge D(y)) \rightarrow (\exists y)(x = h(y) \wedge A(y))). \qquad \text{(Conclusion)}$$

Beginning with the exercises below we shall often omit parentheses when writing predicates. For example, in place of "$A(x)$" we shall write "Ax," and "$A(x,y)$" will be written simply as "Axy."

Exercises

2.6.1. Let Px be the abbreviation for "x is a prime," Ex for "x is even," Ox for "x is odd," and Dxy for "x divides y." Translate each of the following into English.

(a) $P7$.

(b) $E2 \wedge P2$.

(c) $(x)(D2x \rightarrow Ex)$.

(d) $(\exists x)(Ex \wedge Dx6)$.

(e) $(x)(\sim Ex \rightarrow \sim D2x)$.

(f) $(x)(Ex \wedge (y)(Dxy \rightarrow Ey))$.

(g) $(x)(Px \rightarrow (\exists y)(Ey \wedge Dxy))$.

(h) $(x)(Ox \rightarrow (y)(Py \rightarrow \sim Dxy))$.

(i) $(\exists x)(Ex \wedge Px) \wedge \sim(\exists x)((Ex \wedge Px) \wedge (\exists y)(x \neq y \wedge Ey \wedge Py))$.

2.6.2. Below are twenty sentences in English followed by the same number of sentences in symbolic form. Try to pair the members of the two sets in such a way that each member of a pair is a translation of the other member of the pair.

(a) All judges are lawyers. (Jx,Lx)

(b) Some lawyers are shysters. (Sx)

(c) No judge is a shyster.

(d) Some judges are old but vigorous. (Ox,Vx)

(e) Judge Jones is neither old nor vigorous. (j)

(f) Not all lawyers are judges.

(g) Some lawyers who are politicians are Congressmen. (*Px,Cx*)

(h) No Congressman is not vigorous.

(i) All Congressmen who are old are lawyers.

(j) Some women are both lawyers and Congressmen. (*Wx*)

(k) No woman is both a politician and a housewife. (*Hx*)

(l) There are some women lawyers who are housewives.

(m) All women who are alwyers admire some judge. (*Axy*)

(n) Some lawyers admire only judges.

(o) Some lawyers admire women.

(p) Some shysters admire no lawyer.

(q) Judge Jones does not admire any shyster.

(r) There are both lawyers and shysters who admire Judge Jones.

(s) Only judges admire judges.

(t) All judges admire only judges.

(a)′ $(\exists x)(Wx \wedge Cx \wedge Lx)$.

(b)′ $\sim Oj \wedge \sim Vj$.

(c)′ $(x)(Jx \rightarrow \sim Sx)$.

(d)′ $(\exists x)(Wx \wedge Lx \wedge Hx)$.

(e)′ $(x)(Ajx \rightarrow \sim Sx)$.

(f)′ $(x)(Jx \rightarrow Lx)$.

(g)′ $\sim(x)(Lx \rightarrow Jx)$.

(h) $(x)(Cx \wedge Ox \rightarrow Lx)$.

(i)′ $(\exists x)(Lx \wedge Sx)$.

(j)′ $(\exists x)(Lx \wedge Px \wedge Cx)$.

(k)′ $(x)(Wx \rightarrow \sim(Px \wedge Hx))$.

(l)′ $(x)(Cx \rightarrow Vx)$.

(m)′ $(\exists x)(Jx \wedge Ox \wedge Vx)$.

(n)′ $(x)(y)(Ayx \wedge Jx \rightarrow Jy)$.

(o)′ $(\exists x)(Sx \wedge (y)(Axy \rightarrow \sim Ly))$.

(p)′ $(\exists x)(\exists y)(Lx \wedge Sy \wedge Axj \wedge Ayj)$.

(q)′ $(x)(Wx \wedge Lx \rightarrow (\exists y)(Jy \wedge Axy))$.

(r)′ $(\exists x)(Lx \wedge (\exists y)(Wy \wedge Axy))$.

(s)′ $(x)(Jx \rightarrow (y)(Axy \rightarrow Jy))$.

(t)′ $(\exists x)(Lx \wedge (y)(Axy \rightarrow Jy))$.

2.6.3. Using the letters indicated for predicate forms, and whatever symbols of arithmetic (for example, "+" and "<") may be needed, translate the following.

(a) If the product of a finite number of factors is equal to zero, then at least one of the factors is equal to zero. (*Px* for "*x* is a product of a finite number of factors," and *Fxy* for "*x* is a factor of *y*.")

(b) Every common divisor of *a* and *b* divides their greatest common divisor. (*Fxy* for "*x* is a factor of *y*," and *Gxyz* for "*z* is the greatest common divisor of *x* and *y*.")

(c) For each real number *x* there is a larger real number *y*. (*Rx*)

(d) There exist real numbers *x*, *y*, and *z* such that the sum of *x* and *y* is greater than the product of *x* and *z*.

(e) For every real number *x* there exists a *y* such that for every *z*, if the sum of *z* and 1 is less than *y*, then the sum of *x* and 2 is less than 4.

§2.7. First-Order Logic. A Formulation

The examples and exercises in the preceding section serve to substantiate the contention that if the sentential connectives are supplemented with terms, predicates, and quantifiers, much of everyday language can be symbolized unambiguously. In this section we shall formulate languages having the formal counterparts of these notions, along with those for the sentential connectives, as building blocks. The presentation parallels, and is an extension of, our endeavors in §2.3, where a formal language, based on statement letters and the symbols $\sim$, $\vee$, $\wedge$, $\rightarrow$, $\leftrightarrow$, was formulated.* Before we begin it, let us review the pattern of the earlier construction so that it will be clearly fixed in our minds.

There we commenced by specifying the alphabet (statement letters plus the symbols $\sim$, $\wedge$, $\vee$, $\rightarrow$, $\leftrightarrow$) of the language and the finite sequences of these symbols that would constitute grammatically correct sentences (the statement forms) of the language.† Then we gave the semantics of the language. This consisted in specifying the admissible interpretations of statement letters (either T or F) and then providing rules (the truth tables for $\sim$, $\wedge$, $\vee$, $\rightarrow$, $\leftrightarrow$) for determining the interpretation of a sentence accompanying a given interpretation of its constituent statement letters.

The final remark, preparatory to settling down to our task in this section, is this. We shall formulate and then study an infinite family of

---

* Some clarification in this regard is in order. Strictly speaking, there is one such language for each choice of a set of statement letters. Clearly, languages with the same number of statement letters are indistinguishable. In turn, the dependence on the number of statement letters can be eliminated by assuming that the set of statement letters is **denumerable**, that is, in one-to-one correspondence with the set of positive integers. Thus, after the fact, we assume that this is the case in §2.3.

† This is part of the syntax of the language. A description of the syntax is completed with definitions of the rules of proof for the language; then the definition of theorem can be given. These definitions appear in §§3.4 and 3.5.

languages, known as **first-order languages.** Here the adjective "first-order" is used to distinguish the languages we shall describe from those in which there are predicates having other predicates or functions as arguments, or in which predicate quantifiers or function quantifiers are permitted. The word "restricted" is also used with this meaning. One who wishes to emphasize that predicates are basic ingredients of the languages to be discussed might choose to title the study "the restricted calculus of predicates" or simply "predicate calculus."

The primitive symbols (that is, the alphabet) of a first-order language (or a so-called first-order predicate calculus) are the following: the symbols $\sim$ and $\rightarrow$; the punctuation marks ( ) and , (the comma, although not necessary, is convenient for ease in reading formulas); a denumerable set of individual variables x_1, x_2, . . . ; a finite or denumerable nonempty set of predicate letters $P_j^n (n \geq 0, j \geq 1)$; a finite or denumerable, possibly empty, set of function letters $f_j^n (n, j \geq 1)$; and a finite or denumerable, possibly empty, set of individual constants a_1, a_2, ... Thus, in a language, L, some or all of the function letters and individual constants may be absent, and some, but not all, of the predicate letters may be absent. The superscript of a predicate letter indicates the number of arguments assigned to it. The predicate letter P_j^n is called an n-place predicate letter; by admitting the value 0 for n, we can include the logic of statements in our account. The positive integer that is a superscript of a function letter plays a similar role. Usually such superscripts will be omitted if no ambiguity results. The subscript of a predicate letter or a function letter is just an index to distinguish different predicates or function letters with the same number of arguments. If no confusion can result, subscripts will be suppressed.

We proceed next with several definitions for a first-order language L. The notion of **term** is given via the following inductive definition.

a. Individual variables and individual constants are terms.

b. If $t_1, t_2, \ldots, t_n$ are terms, then $f_i^n(t_1, t_2, \ldots, t_n)$ is a term for each function letter f_i^n.

c. An expression is a term only if it can be shown to be a term on the basis of clauses **a** and **b**.

The predicate letters combined with terms produce the **prime formulas** of L: If $t_1, t_2, \ldots, t_n$ are terms, then $P_i^n(t_1, t_2, \ldots, t_n)$ is a prime formula for each predicate letter P_i^n. The **well-formed formulas** (wfs) are defined as follows.

a. A prime formula is a wf.

b. If A and B are wfs and x is a variable, then $(\sim A)$, $(A \rightarrow B)$, and $((x)A)$ are wfs.*

c. Any finite sequence of symbols of L is a wf only if it can be shown to be a wf on the basis of clauses **a** and **b**.

A **subformula** of a wf A is a consecutive part of A that is itself a wf. In $(x)A$, A is called the scope of the (universal) quantifier (x). We introduce other connectives by definitions (see Exercise 2.3.11). If A and B are wfs, then

$$(A \wedge B) \text{ stands for } (\sim(A \rightarrow (\sim B))),$$
$$(A \vee B) \text{ stands for } ((\sim A) \rightarrow B),$$
$$(A \leftrightarrow B) \text{ stands for } ((A \rightarrow B) \wedge (B \rightarrow A)).$$

Further, we define existential quantification as follows:

$$(\exists x)A \text{ stands for } \sim((x)(\sim A)).\dagger$$

The same agreements made in §2.3 regarding the omission of parentheses are made for wfs with the added convention that the quantifiers (x) and $(\exists x)$ apply, as does $\sim$, to the smallest wf following each. For example, $(\exists x)A \vee B$ stands for $((\exists x)A) \vee B$. Finally, we agree to omit parentheses around quantified formulas when they are preceded by other quantifiers. For example, $(x_1)(\exists x_2)(x_3)P(x_1,x_2,x_3)$ stands for $(x_1)((\exists x_2)((x_3)P(x_1,x_2,x_3)))$. Below are further examples to illustrate the omission of parentheses in accordance with our conventions and the inclusion of parentheses to specify the scope of the quantifier "(x_1)." The scope is indicated by the underline.

$(x_1) \underline{P_1(x_1)} \wedge P_2(x_1);$
$(\exists x_2)(x_1)\underline{(P_1(x_1,x_2) \rightarrow (x_3) \, P_2(x_3))};$
$(x_1)\underline{(x_2)(P_1(x_1,x_2) \wedge P_2(x_2,x_3))} \wedge (\exists x_1) \, P_1(x_1,x_2);$
$(x_1)\underline{(P_1(x_1) \wedge (\exists x_1) \, P_1(x_1,x_3) \rightarrow (\exists x_2) \, P_2(x_1,x_2))} \vee P_1(x_1,x_2).\ddagger$

We formulate next several specific formal languages. The genesis of each is a particular mathematical theory. The goal set for the associated

---

* We shall use x, y, z, ... to represent arbitrary variables and A, B, C, ... to represent arbitrary formulas. We have already used t_1, t_2, ..., t_n similarly.

† The basis for this definition is the observation made in Example 2.6.4.

‡ The reader is encouraged to adopt an informal style in writing wfs. This includes using x, y, z, ... to represent arbitrary variables, P, Q, R, ... to represent predicate letters, and omitting commas and parentheses when possible. Then, for example, the last of the displayed wfs might be written as

$$(x)(Px \wedge \exists x Qxz \rightarrow \exists y Rxy) \vee Qxy.$$

language is that it be possible to translate sentences of the theory which may involve quantification over individual variables into wfs of the language. This activity accompanies in a natural way the formulation of a mathematical theory as an axiomatic theory, a topic which is discussed in some detail in Chapter 3. Let it suffice for now to say that such an endeavor entails the adoption of some notions of a theory, along with assumptions about those notions, as basic, and then an attempt to establish that in combination they serve as a foundation for the theory. Hopefully, each example below is carried to the point where a reader gains some assurance that the goal described above for the accompanying language is achieved. We emphasize that at this time our only concern is to persuade the reader, by examples, that formal languages are rich enough to provide formal analogues of the linguistic aspects of mathematical theories.

Examples

2.7.1. Suppose that a practitioner of the axiomatic method were to set out to reconstruct the set theory of Chapter 1. After analyzing how the subject matter was developed, he might conclude that all concepts stem from the membership relation, that is, the 2-place predicate "x is a member of y." This could motivate him to form a language with neither individual constants nor function letters and a single predicate letter P^2. He would abbreviate $P^2(x,y)$ by $x \in y$ and $\sim P^2(x,y)$ by $x \notin y$. Then he might make the following definitions.

$$x = y \quad \text{for } (z)(z \in x \leftrightarrow z \in y);$$
$$x \neq y \quad \text{for } \sim(x = y);$$
$$x \subseteq y \quad \text{for } (z)(z \in x \rightarrow z \in y);$$
$$x \subset y \quad \text{for } x \subseteq y \wedge x \neq y.$$

The next step would be the introduction of some axioms. One might be the sentence "For any sets x and y, there is a set z having x and y as its only members." Its translation into the language described is the following wf.

$$(x)(y)(\exists z)(u)(u \in z \leftrightarrow u = x \vee u = y).$$

2.7.2. As every high-school student knows, the basic ingredients of elementary plane geometry are "points," "lines," and the relation of incidence, "_____ lies on _____." To formulate an axiomatic theory intended to have intuitive geometry as an interpretation, we might use the formal language with two 1-place predicate letters P^1_1 and P^1_2, one 2-place predicate letter P^2, and no individual constants nor function letters. We abbreviate $P^1_1(x)$ by $P(x)$ ("x is a point"), $P^1_2(x)$ by $L(x)$ ("x is a line"), and $P^2(x,y)$ by $I(x,y)$ ("x is on y"). Among the axioms might appear such wfs as

$$(\exists x)P(x), \qquad (\exists x)L(x),$$
$$(x)(y)(I(x,y) \leftrightarrow I(y,x)),$$
$$(x)(P(x) \rightarrow (\exists y)(L(y) \wedge I(x,y))).$$

2.7.3. In Example 3.2.1 appears a formulation of group theory. The following language is adequate for much of this theory. It has one 2-place predicate letter P^2, one function letter f^2 of two arguments, and one individual constant a_1. To conform with the notation used in that example, we shall write $s = t$ for $P^2(s,t)$, $s \cdot t$ for $f^2(s,t)$, and e for a_1. The axioms are expressed by the following wfs.

$$(x)(y)(z)(x \cdot (y \cdot z) = (x \cdot y) \cdot z);$$
$$(x)(x \cdot e = x \wedge e \cdot x = x);$$
$$(x)(\exists y)(x \cdot y = e \wedge y \cdot x = e);$$
$$(x)(x = x);$$
$$(x)(y)(x = y \rightarrow y = x);$$
$$(x)(y)(z)(x = y \wedge y = z \rightarrow x = z);$$
$$(x)(y)(z)(y = z \rightarrow (x \cdot y = x \cdot z \wedge y \cdot x = z \cdot x)).$$

Collectively, the fourth, fifth, and sixth axioms express that $=$ is an equivalence relation and the last expresses that "equals may be substituted for equals" in products.

2.7.4. We propose the following language for the theory of partially ordered sets as described in §1.10. It has two 2-place predicate letters P_1^2 and P_2^2. To conform with usual mathematical notation, we write $x = y$ for $P_1^2(x,y)$ and $x < y$ for $P_2^2(x,y)$. We adopt the following wfs as axioms.

$$(x)(x = x);$$
$$(x)(y)(x = y \rightarrow y = x);$$
$$(x)(y)(z)(x = y \wedge y = z \rightarrow x = z);$$
$$(x)(y)(z)(y = z \rightarrow ((x < y \rightarrow x < z) \wedge (y < x \rightarrow z < x)));$$
$$(x) \sim (x < x);$$
$$(x)(y)(z)(x < y \wedge y < z \rightarrow x < z).$$

Here the first four axioms play the role for this theory that the last four in the preceding example play for group theory. The remaining two establish $<$ as an ordering relation.

We conclude this section with several definitions that pertain to variables. An occurrence of a variable in a wf is **bound** iff its occurrence is within the scope of a quantifier employing that variable or is the explicit occurrence in that quantifier. An occurrence of a variable is **free** if this occurrence of the variable is not bound. For example, in

$$(x_1)P(x_1,x_2)$$

both occurrences of x_1 are bound, and the single occurrence of x_2 is free. Again, in the wf

$$(\exists x_2)(x_1)(P(x_1,x_2) \rightarrow (x_3)\,P(x_3))$$

each occurrence of every variable is bound. A variable is **free in a formula** iff at least one occurrence of it is free, and a variable is **bound in a formula** iff at least one occurrence of it is bound. A variable may be both free and bound in a formula. This is true of x_3 in the wf

$$(x_3)(P(x_3) \wedge (\exists x_1)P_1(x_1,x_3) \rightarrow (\exists x_2)P_2(x_3,x_2)) \vee P_1(x_3,x_1).$$

A wf without free variables is called a **statement*** or a **closed wf.**

Exercises

2.7.1. List the bound and the free occurrences of each variable in each of the following formulas.

(a) $(x)P(x)$.   (d) $(\exists x)P_1(x) \wedge P_2(x)$.
(b) $(x)P(x) \rightarrow P(y)$.   (e) $(\exists x)(y)(P_1(x) \wedge P_2(x)) \rightarrow (x)P_3(x)$.
(c) $P_1(x) \rightarrow (\exists x)P_2(x)$ (f) $(\exists x)(\exists y)(P(x,y) \wedge P(z))$.

2.7.2. An alternative formulation of group theory (see Example 2.7.3) begins with the definition of a group as a nonempty set G together with a binary operation $\cdot$ in G which is associative and such that for given x, y in G, each of the equations $x \cdot z = y$ and $z \cdot x = y$ has a solution in G. A language for symbolizing this definition requires, like that of the example, one 2-place predicate letter (for the equality relation) and one 2-place function letter (for the binary operation) but no individual constant. In place of a 2-place function letter, a 3-place predicate letter can be used. We proceed to substantiate our claims. Let L be the language having just one 2-place predicate letter P^2 and one 3-place predicate letter P^3. We shall write $s = t$ for $P^2(s,t)$ and $P(x,y,z)$ for $P^3(x,y,z)$. As axioms for group theory we take the following wfs.

$(x)(x = x)$.
$(x)(y)(x = y \rightarrow y = x)$.
$(x)(y)(z)(x = y \wedge y = z \rightarrow x = z)$.
$(u)(v)(w)(x)(y)(z)(P(u,v,w) \wedge u = x \wedge v = y \wedge w = z \rightarrow P(x,y,z))$
$(x)(y)(\exists z)P(x,y,z)$
$(x)(y)(z)(w)(P(x,y,z) \wedge P(x,y,w) \rightarrow z = w)$.
$(u)(v)(w)(x)(y)(z)(P(u,v,x) \wedge P(x,w,y) \wedge P(v,w,z) \rightarrow P(u,z,y))$.
$(x)(y)(\exists z)P(x,z,y)$.
$(x)(y)(\exists z)P(z,x,y)$.

Write a paragraph in support of the contention that, collectively, these wfs do serve to define the theory of groups.

* This definition will be found to be consistent with our earlier usage of this term.

§2.8. First-Order Logic. Interpretations and Validity

The semantics of the formal languages described in the preceding section will be formulated in stages, the first of which is the definition of admissible interpretations of wfs. An **interpretation** of a set X of wfs is an ordered pair $\langle D,g \rangle$, where D is a nonempty set, called the **domain** of the interpretation, and g is a mapping having the following attributes. Its domain is the union of the set of predicate letters, the set of function letters, and the set of individual constants, that occur in at least one wf of X. The values of g are defined as follows.

a. $g(P_j^n)$ is an n-ary relation in D for an n-place predicate letter P_j^n. In particular, g assigns to a 2-place predicate letter reserved for equality the identity relation $=$ in D, defined by $a = b$ iff a and b are the same.

b. $g(f_j^n)$ is an n-ary operation in D, that is, a function of D^n into D, for an n-place function letter f_j^n.

c. $g(a_i)$ is a fixed element of D for an individual constant a_i.

We offer next several examples of interpretations of sets of wfs. In these, variables are thought of as ranging over the domain of the interpretation, and sentential connectives and quantifiers are given their intuitive meaning. Since, thereby, each wf acquires meaning, the question of its truth or falsity can be raised. Our intuitive discussion in this connection provides guidelines for the presentation following the examples of a formal definition of *truth* and *falsity* for wfs.

Examples

2.8.1. In Example 2.7.3 appears a set of axioms for group theory. Let $\langle D,g \rangle$ be the interpretation of this set of wfs with $D = \{a,b\}$, $g(\cdot) = \circ$ (the operation defined below), and $g(e) = a$.

$\circ$	a	b
a	a	b
b	b	a

The statement $a \circ (b \circ a) = (a \circ b) \circ a$ is true because $a \circ (b \circ a) = a \circ b = b$ and $(a \circ b) \circ a = b \circ a = b$. Similarly, each of the remaining seven statements of the form $x \circ (y \circ z) = (x \circ y) \circ z$ can be shown to be true. Hence $(x)(y)(z)(x \cdot (y \cdot z) = (x \cdot y) \cdot z)$ is true for $\langle D,g \rangle$. Likewise the axiom $(x)(x \cdot e = x \wedge e \cdot x = x)$ is true because $a \circ a = a$ and $a \circ b = b \circ a = b$. Similarly the remaining axioms are found to be true. That the axioms of group theory are true for the interpretation $\langle D,g \rangle$ will also be expressed by stating that $\langle D,g \rangle$ is a *model* for (the axioms of) group theory.

2.8.2. Let L be the language proposed in Example 2.7.4 for the theory of partially ordered sets. Let $\langle D,g \rangle$ be the interpretations of its axioms where $D = \{a,b\}$ and $g(<) = \{\langle a,b \rangle\}$. It is easily verified that $\langle D,g \rangle$ is a model of the theory.

Consider next the interpretation $\langle \mathbb{R},g \rangle$ where $g(<)$ is the familiar relation "is less than" in the set of real numbers. We ask whether the statement

$$A: (x)(y)(x < y \rightarrow (\exists z)(x < z \wedge z < y))$$

is true or false. Assign to x a real number r and to y a real number s. If r is not less than s, the conditional is true. If r is less than s, there does exist an assignment to z, for example $(r + s)/2$, such that the consequent and hence the conditional are true. We conclude that A is true for $\langle \mathbb{R},g \rangle$.

On the other hand, A is easily shown to be false for the interpretation $\langle \mathbb{Z},g \rangle$ where $g(<)$ is the relation "is less than" for integers.

2.8.3. As is illustrated above, the translation of a statement (that is, a wf without free variables) is either a true or a false sentence. The translation of a wf with free variables is a relation on the domain of the interpretation which may be true for some values in the domain of the free variables and false for the others. As an example, consider the following three wfs:

$$\text{(i) } P(x_1, f(x_2));$$
$$\text{(ii) } (x_2)P(x_1, f(x_2));$$
$$\text{(iii) } (x_1)(x_2)P(x_1, f(x_2))$$

Let $\langle \mathbb{R},g \rangle$ be the interpretation such that $g(P) = <$, and $g(f): u \mapsto u^2$. Then the translation of (i) is the relation $\{\langle u,v \rangle | u < v^2\}$. If a and b are assigned to x_1 and x_2 respectively, then $P(x_1, f(x_2))$ is true if $a < b^2$ and otherwise is false. The translation of (ii) is the property "is less than the square of every real number"; thus (ii) is true if to x_1 is assigned a negative real number and otherwise is false. Finally, the statement (iii) is false for the interpretation.

Suppose now that L is a first-order language and that $I = \langle D,g \rangle$ is an interpretation of (the wfs of) L. At the core of the assignment of truth values (relative to I) to wfs of L is the assignment of values in D to the variables which appear in one or another wf. To cope with this matter, it is expeditious to make simultaneous assignments to the set $V = \{x_1, x_2, \ldots, x_n, \ldots\}$ of *all* individual variables of L. Any such assignment is simply a function s on V into D. In the first part of the formulation of an algorithm for assigning truth values to wfs, these functions play a dominant role. This is a reflection of the importance that the assignments to variables has in the determination of truth values.

A function $s: V \to D$ may be extended to a function $\bar{s}$ on the set of terms of L into D by the following recursive definition.

   a.   For an individual variable x_i, $\bar{s}(x_i) = s(x_i)$.
   b.   For an individual constant a_i, $\bar{s}(a_i) = g(a_i)$
   c.   For a function letter f_j^n and terms $t_1, \ldots, t_n$,

$$\bar{s}(f_j^n(t_1, \ldots, t_n)) = g(f_j^n)(\bar{s}(t_1), \ldots, \bar{s}(t_n)).$$

Note that this is an extension of s relative to I. Intuitively, $\bar{s}(t)$, for a term t, is the element of D obtained by substituting $g(a)$ for a constant a appearing in t, substituting $s(x)$ for all occurrences of a variable x in t, and then performing the operations assigned by I to the function letters of t. For example, if t is $f_1^2(x_4, f_2^2(a_1, x_1))$ and $I = \langle \mathbb{Z}, g \rangle$ where $g(f_1^2) = +$, $g(f_2^2) = \cdot$, and $g(a_1) = 2$, then $\bar{s}(t)$ is the integer $s(x_4) + 2 \cdot s(x_1)$.

Next we shall define what it means for I to **satisfy** (a wf) A **with** s, symbolized

$$\vDash_I A[s].$$

Intuitively, $\vDash_I A[s]$ iff the translation of A determined by I is true when each free occurrence in A of a variable x is translated as $s(x)$. The formal definition is given inductively as follows.

   a.   If A is an atomic wf $P_j^n(t_1, \ldots, t_n)$, $\vDash_I P_j^n(t_1, \ldots, t_n)[s]$ iff $\langle \bar{s}(t_1), \ldots, \bar{s}(t_n) \rangle \in g(P_j^n)$.

Remark

   Some illustrations may be helpful. Suppose A is $P(f(x_2, x_4), x_1)$ and $I = \langle \mathbb{R}, g \rangle$, where $g(P) = \leq$ and $g(f) = +$. Then $\vDash_I A[s]$ iff $s(x_2) + s(x_4) \leq s(x_1)$. Next let A be

$$P(x_4, f(x_1, x_2), f(x_1, x_3)).$$

Suppose $I = \langle D, g \rangle$ where D is the set of points in the plane, $g(P)$ is the ternary relation of all triples $\langle p, q, r \rangle$ of points such that p is closer to q than to r, and $g(f)$ is the function which assigns to $\langle p, q \rangle$ the midpoint of $\overline{pq}$, the line segment joining p and q. Then $\vDash_I A[s]$ iff $s(x_4)$ is closer to the midpoint of $\overline{s(x_1)s(x_2)}$ than to the midpoint of $\overline{s(x_1)s(x_3)}$.

Notice that in view of the agreement made in clause **a** of the definition of an interpretation, if P_1^2, for example, is reserved for equality, then $\vDash_I P_1^2(t_1, t_2)[s]$ iff $\bar{s}(t_1) = \bar{s}(t_2)$.

b. $\models_I (\sim A)[s]$ iff not $\models_I A[s]$.

c. $\models_I (A \to B)$ $[s]$ iff either not $\models_I A[s]$ or $\models_I B[s]$.*

d. $\models_I (x_i)A[s]$ iff for every $d \in D$, $\models_I A[s(x_i|d)]$, where $s(x_i|d)$ is the modification of s that results upon assigning the value d to the variable x_i.

From the definitions of $\wedge$, $\vee$, $\leftrightarrow$, and $\exists$, we find immediately that

e. $\models_I (A \wedge B)[s]$ iff $\models_I A[s]$ and $\models_I B[s]$;

f. $\models_I (A \vee B)[s]$ iff $\models_I A[s]$ or $\models_I B[s]$;

g. $\models_I (A \leftrightarrow B)[s]$ iff $\models_I A[s]$ and $\models_I B[s]$ or not $\models_I A[s]$ and not $\models_I B[s]$;

h. $\models_I (\exists x_i)A[s]$ iff $\models_I A[S(x_i|d)]$ for some d in D.

We establish next a fact which should be obvious intuitively, namely, that satisfaction of a wf A with a given s depends only on the values of s at the (finitely many) variables which occur free in A.

THEOREM 2.8.1. Let $I = \langle D, g \rangle$ be an interpretation of the wf A and let s_1, s_2 be functions on V into D which agree at all variables (if any) which occur free in A. Then $\models_I A[s_1]$ iff $\models_I A[s_2]$.

Proof. Assume s_1 and s_2 satisfy the condition stated. We show first that if A is a prime wf, then $\bar{s}_1(t) = \bar{s}_2(t)$ for every term t of A. If t is a variable, then it occurs free in A and hence $\bar{s}_1(t) = s_1(t) = s_2(t) = \bar{s}_2(t)$. If t is a constant, then $\bar{s}_1(t) = s_1(t) = g(t) = s_2(t) = \bar{s}_2(t)$. Suppose that

$$\bar{s}_1(t) = \bar{s}_2(t_1), \ldots, \bar{s}_1(t_m) = \bar{s}_2(t_m)$$

and t is $f(t_1, \ldots, t_m)$, where f is an m-place function letter. Let $g(f) = F$. Then

$$\bar{s}_1(t) = F(\bar{s}_1(t_1), \ldots, \bar{s}_1(t_m)) = F(\bar{s}_2(t_1), \ldots, \bar{s}_2(t_m)) = \bar{s}_2(t).$$

We now prove the theorem by induction on the number n of symbols in A, counting each occurrence of $\sim$, $\to$, and a universal quantifier as a symbol. If $n = 0$, then A is a prime formula $P(t_1, \ldots, t_m)$. Thus $\models_I A[s_1]$ iff $\langle \bar{s}_1(t_1), \ldots, \bar{s}_1(t_m) \rangle \in g(P)$ and $\models_I A[s_2]$ iff $\langle \bar{s}_2(t_1), \ldots, \bar{s}_2(t_m) \rangle \in g(P)$. Since $\bar{s}_1(t_i) = \bar{s}_2(t_i)$, $1 \leq i \leq m$, I satisfies A with s_1 iff it satisfies A with s_2.

Assume the theorem holds for every wf with n or fewer symbols and consider A with $n + 1$ symbols. If A is $\sim B$ or $B \to C$ for some B and

---

* In other words, if I satisfies A with s, then I satisfies B with s.

C, the proof is immediate by the induction hypothesis and the definition of $\models_I A[s]$. There remains the case where A is $(x)B$ for some x
and B. Then the variables free in A are those free in B with the ex-
ception of x. Thus for each d in D, $s_1(x|d)$ and $s_2(x|d)$ agree at all
variables free in B. By the induction hypothesis, I satisfies B with
$s_1(x|d)$ iff it satisfies B with $s_2(x|d)$. From this and the definition of
satisfaction it follows that I satisfies A with s_1 iff it does so with s_2.

COROLLARY. If A is a statement, either I satisfies A with every s:
$V \rightarrow D$ or I satisfies A with no s.

A wf A of L is **true for the interpretation I** = $\langle D,g \rangle$ iff I satisfies A
with every s and A is **false for I** iff I does not satisfy A with any s. If
A is a statement, then the corollary states that either A is true for I or A
is false for I. (They cannot both hold, since D is nonempty.)* If a wf A
is true for I, then I is a **model** of A; I is a model of a set X of wfs iff it
is a model of each wf in X.

In the next theorem we collect some basic results concerning the
truth and falsity of wfs.

THEOREM 2.8.2. Let I be an interpretation of the wfs A and B.
   (a) A is true for I iff $\sim A$ is false for I; A is false for I iff $\sim A$ is true
      for I.
   (b) A is not both true and false for I.
   (c) If A is a statement, then exactly one of A and $\sim A$ is true for I and
      the other is false for I.
   (d) If $A \rightarrow B$ and A are true for I, so is B.
   (e) $A \rightarrow B$ is false for I iff A is true for I and B is false for I.
   (f) A is true for I iff $(x_i)A$ is true for I.

Proof.
   (a) We prove the first assertion. By definition, A is true for I iff I
      satisfies A with every s. From clause (b) of the definition of satis-
      faction, $\models_I A[s]$ iff not $\models_I (\sim A)[s]$. Hence, A is true for I iff I
      does not satisfy $\sim A$ with any s. The latter holds iff $\sim A$ is false
      for I.
   (b) Suppose A is both true and false for I. Then I satisfies A with

---

* Recall that a statement, as defined within first-order logic, is a wf without free variables.
Thus, a 0-place predicate letter is a statement, and consequently is either true for an interpretation I or false for I. This observation establishes the consistency of our present definition
with that used earlier.

every s and I satisfies A with no s. This is impossible because the domain is nonempty, and so there exists at least one s.

(c) This is a restatement of the corollary to Theorem 2.8.1.

(d) Suppose $A \to B$ and A are true for I. Then I satisfies $A \to B$ and A with every s. From clause (c) of the definition of satisfaction, it follows that I satisfies B with every s.

The proofs of (e) and (f) are left as exercises.

Examples

2.8.4. The **closure** of a wf A is the wf obtained from A by prefixing as universal quantifiers those variables (in order of increasing subscripts) which are free in A. It follows from (f) of Theorem 2.8.2 that A is true for I iff its closure is true for I.

2.8.5. An **instance** of a statement form is a wf obtained from the form by substituting wfs for all statement letters, with all occurrences of the same statement letter being replaced by the same wf. We contend that every instance of a tautology A is true for any interpretation $I = \langle D,g \rangle$. Indeed, suppose A involves n statement letters and that we substitute for them, in some order, the wfs $A_1, A_2, \ldots, A_n$ with all occurrences of the first being replaced by A_1, and so on. Denote the resulting wf by A'. Let $s: V \to D$. Then, for each i, I satisfies A_i with s or I does not satisfy A_i with s. Now construct a table having the format of a truth table for A', regarded as built up from $A_1, A_2, \ldots, A_n$, but writing *Yes* for I *satisfies* A_i *with* s and *No* for I *does not satisfy* A_i *with* s. By the definition of satisfaction, this table translates into the truth table for A upon replacing *Yes* by T and *No* by F. Since A is a tautology, in the *Yes-No* table there must appear in the column under A' only *Yes*. Thus I satisfies A' with s.

2.8.6. We prove that $(x)P(x) \to P(y)$, where P is a 1-place predicate letter, is true for all interpretations. Assume the assertion is incorrect. Then $(x)P(x) \to P(y)$ is not true for some interpretation $I = \langle D,g \rangle$. Suppose $g(P)$ is the subset D' of D. By clause (c) of the definition of satisfaction, there exists an s such that $\models_I (x)P(x)[s]$ and not $\models_I P(y)[s]$. Then $s(y) \not\subseteq D'$ by clause (a) of the definition. By clause (d) of the definition, $\models_I P(x)[s(x|d)]$ for all d in D, that is, $d \in D'$ for all d in D. This with $s(y) \not\subseteq D'$ is a contradiction. Hence the assertion is proved.

2.8.7. We prove that $P(y) \to (\exists x)P(x)$ is true for all interpretations. Let $I = \langle D,g \rangle$ be an interpretation of the wf with $g(P) = D'$. It is sufficient to show that if $\models_I P(y)[s]$, then $\models_I P(x)[s(x|d)]$ for some d in D. Assume $\models_I P(y)[s]$. Then $s(y) \in D'$ and hence $\models_I P(x)[s(x|s(y))]$.

2.8.8. We prove that the wf $(\exists x)P(x) \to (x)P(x)$ is false for $I = \langle D,g \rangle$, where $D = \{a,b\}$ and $g(P) = \{a\}$. Indeed, $\models_I P(x)[s(x|a)]$, so $\models_I (\exists x)P(x)[s]$, but not $\models_I P(x)[s(x|b)]$, so not $\models_I (x)P(x)[s]$.

A wf A of L is (**logically**) **valid** iff A is true for every interpretation. For "A is valid" we shall write

$$\models A.$$

We do so in order to use the same terminology and symbolism introduced earlier for statement forms, since this definition of validity is an extension of the earlier one (see Example 2.8.5). But there is an important difference in the level of complexity. The question of the validity of a statement form can be settled by constructing a truth table; this can be done in a finite number of steps following a recipe. In contrast, to investigate whether a wf A is valid requires a consideration of all possible interpretations $I = \langle D, g \rangle$. For each interpretation, in turn, each function s on V into D must be considered. And for each s, relative to a given I, it must be determined whether I satisfies A with s. With these compounded complications it is not surprising that the set of valid wfs of L is undecidable (see §3.7); hence that reasoning processes cannot be replaced by a recipe. This state of affairs leads in a natural way to a search for general methods to prove the validity of wfs. As an illustration of what we have in mind, we might ask: Having proved above that $\models (x)P(x) \to P(y)$, is there a method for inferring the validity of more complex wfs which have "the same form"? Even in the statement calculus, where there is no necessity *in principle* for such methods, we offered some simply to expedite the determination of tautologies.

Thus we begin by looking for what we can take over from the statement calculus.* Let us call the wfs A and B (**logically**) **equivalent,** symbolized

$$A \text{ eq } B,$$

iff for every interpretation I, if I satisfies A with s then I satisfies B with s and conversely. This is the analogue for L of the eq relation for statement forms. It is easily shown that there is the following analogue of Theorem 2.3.2.

THEOREM 2.3.2′. For wfs A and B, $\models A \leftrightarrow B$ iff A eq B.

We shall content ourselves with the following extension of the corollary to Theorem 2.3.2.

---

* This investigation proves to be unrewarding in terms of its intended goal. However, it does turn up some interesting notions which pertain to wfs.

COROLLARY. Let C_A be a wf containing a specific occurrence of the wf A and let C_B be the result of replacing that occurrence of A by a wf B. If $A \leftrightarrow B$ is an instance of a tautology, then $\vDash C_A \leftrightarrow C_B$. If, in addition, $\vDash C_A$, then $\vDash C_B$.

Part (d) of Theorem 2.8.2 yields a generalization of Theorem 2.3.3.

THEOREM 2.3.3′. If $\vDash A$ and $\vDash A \rightarrow B$ for wfs A and B, then $\vDash B$.

Theorem 2.3.1 has various extensions to wfs. Each involves restrictive conditions to avoid the binding, in a way which is not intended, of variables which may be present. In order to give an illustration, some notational conventions will be introduced. We shall often indicate that in a wf A some of the variables $x_1, x_2, \ldots, x_n$ (for example) have free occurrences by writing it as $A(x_1, x_2, \ldots, x_n)$. Such notation is not to be confused with that used for prime wfs, nor does it mean that these variables necessarily have free occurrences in A, nor that other variables do not have free occurrences in A. This notation is convenient because we can then agree to write $A(t_1, t_2, \ldots, t_n)$ for the result of substituting in A the terms $t_1, t_2, \ldots, t_n$ for all free occurrences (if any) of $x_1, x_2, \ldots, x_n$, respectively.

Recall that in §2.3 it was pointed out that a statement form which has the "same form" as a tautology is a tautology. This may suggest, with wfs in mind, that since $\vDash (x)P(x) \rightarrow P(y)$, it follows that $\vDash (x)A(x) \rightarrow A(y)$ for any wf A. This is not true. For example, if $A(x)$ is $(\exists y)P(x,y)$, then $(x)A(x) \rightarrow A(y)$ is

$$(x)(\exists y)P(x,y) \rightarrow (\exists y)P(y,y)$$

which is false for the interpretation having $\mathbb{Z}$ as domain and that assigns to P the relation "is less than." The breakdown stems from the fact that, upon replacing x by y in $P(x,y)$, the resulting occurrence of y is bound by $\exists y$. Mishaps of this kind can arise when one merely substitutes one variable for another in a wf. As an illustration, let $A(x)$ be $(\exists y)(y > x)$, a wf of the language for elementary arithmetic. Then

$$A(2) \quad \text{is} \quad (\exists y)(y > 2),$$
$$A(z) \quad \text{is} \quad (\exists y)(y > z),$$
$$A(y) \quad \text{is} \quad (\exists y)(y > y).$$

Whereas the first two wfs reflect the intended meaning of $A(x)$, the last does not, because the y which has been substituted for x falls within the scope of $\exists y$ and becomes bound. As a result, $A(y)$ does not "say the same thing" about y as $A(x)$ does about x. In everyday mathematics one is not likely to make a substitution which changes the meaning of a formula. A safeguard against inappropriate substitutions in formal situations can be given. Let x and y be variables and $A(x)$ a wf. Then $A(x)$ **admits** y **for** x iff every free occurrence of x in $A(x)$ becomes a free occurrence of y in $A(y)$. More generally, $A(x)$ **admits** t **for** x iff $A(x)$ admits y for x for every variable y that occurs in the term t.

Some examples are in order. If $A(x)$ is $(\exists y)(y > x)$, then $A(x)$ does not admit y for x. It does admit every variable other than y for x. The wf $(y)P_1(x,y) \vee (\exists z)P_2(y,z)$ admits z for x but does not admit the term $f(z,y)$ for x. The wf $(y)P_1(x,z) \rightarrow P_2(x)$ admits $f(x,z)$ for x, but $(\exists z)(y)P_1(x,y) \rightarrow P_2(x)$ does not admit $f(x,z)$ for x.

We are now in a position to state the following extensions of the results in Examples 2.8.6 and 2.8.7.

THEOREM 2.8.3. If $A(x_i)$ is a wf that admits the term t for x_i, then

    (I) $\vDash (x_i)A(x_i) \rightarrow A(t)$;

    (II) $\vDash A(t) \rightarrow (\exists x_i)A(x_i)$.

Proof. The proof of (I) requires considerable preparation and will not be given. Accepting (I), a proof of (II) can be given as follows. Assume the wf in (II) is not valid. Then it is not true for some interpretation I. Hence there exists an s such that I satisfies $A(t)$ with s but does not satisfy $\sim(x_i)(\sim A(x_i))$ with s, that is, I satisfies $A(t)$ and $(x_i)(\sim A(x_i))$ with s. By (I), $(x_i)(\sim A(x_i)) \rightarrow \sim A(t)$ is valid, and so this wf is true for I. It follows that I satisfies it with s. Hence, since I satisfies $(x_i)(\sim A(x_i))$ with s, I satisfies $\sim A(t)$ with s. Thus our assumption leads to a contradiction, and so the wf in question is valid.

COROLLARY. For any wf $A(x_i)$, if $\vDash (x_i)A(x_i)$ then $\vDash A(x_i)$.

Proof. If $A(x_i)$ is any wf, it admits x_i for x_i. Thus we may apply (I) of the theorem to obtain $\vDash (x_i)A(x_i) \rightarrow A(x_i)$. Now assume that $\vDash (x_i)A(x_i)$. Then $\vDash A(x_i)$ by Theorem 2.3.3'.

We offer next another pair of results which will also find application later.

THEOREM 2.8.4. Let A be any wf and B be any wf which does not contain a free occurrence of the variable x_i.

    (I) If $\vDash B \rightarrow A$, then $\vDash B \rightarrow (x_i)A$.

    (II) If $\vDash A \rightarrow B$, then $\vDash (\exists x_i)A \rightarrow B$.

Proof. For the proof of (I) we need the following result. If an interpretation $I = \langle D,g \rangle$ satisfies a wf B with s and x_i is not free in B, then I satisfies B with $s(x_i|d)$ for all d in D. For the proof assume that $\vDash_I B[s]$ and x_i is not free in B. Let $x_{j_1}, x_{j_2}, \ldots, x_{j_n}$ be the free variables of B. Then x_i is not one of them. Let d be in D. Then $s(x_i|d)$ agrees with s at $x_{j_1}, x_{j_2}, \ldots, x_{j_n}$. Hence $\vDash_I B[s(x_i|d)]$ by Theorem 2.8.2.

To establish (I), we assume that $\vDash B \to A$ and show that if an interpretation I satisfies B with s, then I satisfies A with $s(x_i|d)$ for all d. Assume that $\vDash_I B[s]$. Then, since B has no free occurrence of x_i, $\vDash_I B[s(x_i|d)]$ for all d. Hence, from our assumption, $\vDash_I A[s(x_i|d)]$ for all d.

The proof of (II) is left as an exercise.

At this point we could undertake a direct generalization of Theorem 2.3.1 which would reduce the proof of the validity of each of the wfs in the next theorem to the case of prime wfs in place of arbitrary wfs. Since this strategy yields no simplification in the proofs, we omit the undertaking and proceed directly to the theorem. We continue the numbering of Theorem 2.3.4 (which holds for wfs) to emphasize that we are extending our list of valid wfs.

THEOREM 2.8.5. Let A and B be any wfs and C be any wf having no free occurrence of the variable x_i. Then

33. $\vDash (x_i)(x_j)A \leftrightarrow (x_j)(x_i)A$.   33′. $\vDash (\exists x_i)(\exists x_j)A \leftrightarrow (\exists x_j)(\exists x_i)A$.
34. $\vDash (x_i)A \leftrightarrow \sim(\exists x_i)\sim A$.   34′. $\vDash (\exists x_i)A \leftrightarrow \sim(x_i)\sim A$.
35. $\vDash \sim(x_i)A \leftrightarrow (\exists x_i)\sim A$.   35′. $\vDash \sim(\exists x_i)A \leftrightarrow (x_i)\sim A$.
36. $\vDash (\exists x_i)(x_j)A \to (x_j)(\exists x_i)A$.
37. $\vDash (x_i)(A \wedge B) \leftrightarrow (x_i)A \wedge (x_i)B$.
  37′. $\vDash (\exists x_i)(A \vee B) \leftrightarrow (\exists x_i)A \vee (\exists x_i)B$.
38. $\vDash (x_i)A \vee (x_i)B \to (x_i)(A \vee B)$.
  38′. $\vDash (\exists x_i)(A \wedge B) \to (\exists x_i)A \wedge (\exists x_i)B$.
39. $\vDash (x_i)(C \wedge B) \leftrightarrow C \wedge (x_i)B$.
  39′. $\vDash (\exists x_i)(C \vee B) \leftrightarrow C \vee (\exists x_i)B$.
40. $\vDash (x_i)(C \vee B) \leftrightarrow C \vee (x_i)B$.
  40′. $\vDash (\exists x_i)(C \wedge B) \leftrightarrow C \wedge (\exists x_i)B$.

Proof. The validity of each of the wfs appearing in the righthand column can be established by formal manipulation from that of its correspondent in the lefthand column. The method employs Theorem 2.3.2′ and its corollary, in conjunction with the definition of an existential quantifier. For example, from 40 we have, upon replacing

C by $\sim C$, and B by $\sim B$, and using the fact that $\models A \leftrightarrow B$ implies $\models \sim A \leftrightarrow \sim B$,

$$\models \sim(x_i)(\sim C \vee \sim B) \leftrightarrow \sim(\sim C \vee (x_i)\sim B).$$

In turn,

$$\models \sim(x_i) \sim (C \wedge B) \leftrightarrow \sim(\sim C \vee \sim(\sim(x_i)\sim B)),$$
$$\models (\exists x_i)(C \wedge B) \leftrightarrow \sim(\sim C \vee \sim(\exists x_i)B),$$
$$\models (\exists x_i)(C \wedge B) \leftrightarrow C \wedge (\exists x_i)B,$$

which is 40′. We also mention that the validity of the wfs in 34 and 35 may be shown by similar manipulations.

The validity of the other wfs in the lefthand column may be established by applying the definition of that concept. As an illustration we prove 40. Assume an interpretation I satisfies $(x_i)(C \vee B)$ with s. Then $\models_I C \vee B[s(x|d)]$ for all d. Then $\models_I C[s(x_i|d)]$ or $\models_I B[s(x_i|d)]$. It follows that $\models_I C \vee (x_i)B[s]$. Conversely, suppose $\models_I C \vee (x_i)B[s]$. Then $\models_I C[s]$ or $\models_I (x_i)B[s]$. In the first case we may infer that $\models_I C[s(x_i|d)]$ by the preliminary result derived in the proof of Theorem 2.8.4. In the second case it follows that $\models_I B[s(x_i|d)]$. Hence, $\models_I C \vee B[s(x_i|d)]$, that is, $\models_I(x_i)(C \vee B)[s]$. Thus, $\models (x_i)(C \vee B) \leftrightarrow C \vee (x_i)B$.

The results in Theorem 2.8.5 provide a variety of valuable working rules in practical applications. Formulas 33 and 33′ justify the interchange of universal quantifiers and of existential quantifiers and 34, 34′ describe how a universal quantifier can be expressed in terms of an existential quantifier and conversely. Formulas 35 and 35′ provide rules for transferring $\sim$ across quantifiers. The others are concerned with the transfer of quantifiers across $\vee$ and $\wedge$.

Examples

2.8.9. We consider some practical illustrations of 35 and 35′ in the arithmetic of natural numbers; thus we assume that variables range over the set $\mathbb{N} = \{0, 1, 2, \ldots\}$. We shall use familiar symbols for predicates (relations) and functions (operations). The (true) statement "There does not exist a greatest natural number" may by symbolized by

$$(x)(\exists y)(x < y).$$

Its negation,

$$\sim(x)(\exists y)(x < y),$$

may be rewritten, using 35′, as

$$(\exists x) \sim ((\exists y)(x < y)).$$

In turn, using 35, this may be rewritten as

$$(\exists x)(y) \sim (x < y) \text{ or } (\exists x)(y)(x \geq y).$$

In English this last formula reads "There exists a greatest natural number."

The (false) statement "For every pair m, n of natural numbers there is a natural number p such that $m + p = n$" may be symbolized by

$$(m)(n)(\exists p)(m + p = n).$$

Its negation may be transformed into

$$(\exists m)(\exists n)(p)(m + p \neq n).$$

The reader can translate this into acceptable English.

2.8.10. In this example we again use familiar symbols for relations and functions. The domain of the interpretation is the set of real numbers. The definition of continuity of a function f at a, namely, "f is continuous at a iff for every $\epsilon > 0$ there exists a $\delta > 0$ such that for all x, if $|x - a| < \delta$, then $|f(x) - f(a)| < \epsilon$" can be translated into the symbolic form

(1) $(\epsilon)(\epsilon > 0 \rightarrow ((\exists \delta)(\delta > 0 \wedge (x)(|x - a| < \delta \rightarrow |f(x) - f(a)| < \epsilon)))$.

A working form for the negation of this wf is

(2) $(\exists \epsilon)(\epsilon > 0 \wedge (\delta)(\delta > 0 \rightarrow (\exists x)(|x - a| < \delta \wedge |f(x) - f(a)| \geq \epsilon)))$.

It may be obtained quickly using instances of the following results, which the reader should verify.

$$\sim(u)(A \rightarrow B) \text{ eq } (\exists u)(A \wedge \sim B).$$
$$\sim(\exists u)(A \wedge B) \text{ eq } (u)(A \rightarrow \sim B).$$

The wf (1) can be shortened using "restricted" quantification. This notion can be explained as follows. A statement of the form "There exists an x of a certain kind such that $A(x)$" may be symbolized as

(3) $(\exists x)(K(x) \wedge A(x)),$

and one of the form "For all x of a certain kind, $A(x)$" may be symbolized as

(4) $(x)(K(x) \rightarrow A(x)).$

If we specify, before symbolizing, that x is restricted to be of this certain kind, then we replace (3) by "$(\exists x)A(x)$" and (4) by "$(x)A(x)$." For example, if we restrict δ and ϵ to be positive numbers, then (1) may be rewritten as

(5) $(\epsilon)(\exists \delta)(x)(|x < a| < \delta \rightarrow |f(x) < f(a)| < \epsilon).$

With mild restrictions, the valid formulas of Theorem 2.8.5 remain valid when some quantifiers are restricted. This makes it possible, for example, to obtain the negation of complicated formulas quickly and in abbreviated form. As an illustration, the negation of (5) is found to be

$$(\exists \epsilon)(\delta)(\exists x)(|x - a| < \delta \wedge |f(x) - f(a)| \geq \epsilon).$$

This is precisely the simplification of (2) that results with the same restrictions applied to the quantifiers.

Exercises

2.8.1. For each of the wfs

(i) $P(f(x_1,x_2), a_1)$,

(ii) $P(x_1,x_2) \to P(x_2,x_1)$,

(iii) $(x_1)(x_2)(x_3)(P(x_1,x_2) \to (P(x_2,x_3) \to P(x_1,x_3)))$,

consider the following interpretations:

$I_1 = \langle D_1, g_1 \rangle$, where $D_1 = \mathbb{Z}^+$, $g_1(P) = \geq$, $g_1(f)$ is multiplication and $g_1(a_1) = 1$;

$I_2 = \langle D_2, g_2 \rangle$, where D_2 is the collection of all sets of integers, $g_2(P) = \supseteq$, $g_2(f)$ is the operation of union, and $g_2(a_1) = \varnothing$.

(a) If I_1 is the interpretation of (i), for what values of x_1 and x_2 is (i) true?

(b) If I_1 is the interpretation of (ii), for what values of x_1 and x_2 is (ii) true?

(c) Is (iii) true or false for I_1?

(d) Answer the above questions with I_2 in place of I_1.

2.8.2. Show that the assertions labeled (e)–(h) following the definition of satisfaction are consequences of that definition.

2.8.3. Supply proofs of parts (e) and (f) of Theorem 2.8.2.

2.8.4. Prove that if the wf A has no free occurrences of x_i, then $\vDash (x_i)(A \to B) \to (A \to (x_i)B)$.

2.8.5. Prove part (II) of Theorem 2.8.4.

2.8.6. Show that the wfs in 34 and 35 of Theorem 2.8.5 are valid.

2.8.7. Show that the wfs in 33, 36, 37, 38, and 39 of Theorem 2.8.5 are valid.

2.8.8. Show that $(x_j)(\exists x_i)A \to (\exists x_i)(x_j)A$ is not valid.

2.8.9. Using Theorem 2.8.5, show that the members of each of the following pairs of wfs of arithmetic are equivalent wfs.

(a) $(\exists x)(y) \sim (y > x)$, $(\exists x) \sim (\exists y)(y > x)$.

(b) $(\exists x)(y)(y > x \lor \sim(y > 0))$, $(\exists x)(y)(y > 0 \to y > x)$.

(c) $(x)(\exists y)(\exists z)(x < y \land z^2 > y)$, $(x)(\exists y)(x < y \land (\exists z)(z^2 > y))$.

2.8.10. Let $a_0, a_1, \ldots, a_n, \ldots$ be a sequence of real numbers. Using restricted quantification, translate into symbolic form

(a) the assertion that a is the limit of the sequence;

(b) the assertion that the sequence has a limit;

(c) the assertion that the sequence is a Cauchy sequence (that is, given $\epsilon > 0$ there exists a positive integer k such that if $n, m > k$, then $|a_n - a_m| < \epsilon$).

2.8.11. Write the negation of each of the formulas obtained in the preceding exercise.

2.8.12. With $\mathbb{R}$ as domain, translate each of the following statements into symbolic form, write the negation of each (transferring $\sim$ past the quantifiers), and translate each resulting formula into English.

(a) For $x, y \in \mathbb{R}$ and $z \in \mathbb{R}^+$, $xz = yz$ implies $x = y$.

(b) The number a is the least upper bound of $A \subseteq \mathbb{R}$.

(c) The set A has a greatest element.

§2.9. First-Order Logic. Consequence

The concept of consequence for first-order logic is an extension of that for the statement calculus as presented in §2.4. In this extension the method for assigning truth values to statement forms gives way to the notion of satisfaction of wfs by interpretations. The basic definition is the following. A wf B of a first-order language L is a (**logical**) **consequence** of a set X of wfs of L, symbolized

$$X \vDash B,$$

iff for every interpretation $I = \langle D,g \rangle$ of L and every function $s: V \to D$ such that I satisfies every member of X with s, I also satisfies B with s.*
If $X = \{A_1, A_2, \ldots, A_m\}$, we shall write $A_1, A_2, \ldots, A_m \vDash B$ in place of $\{A_1, A_2, \ldots, A_m\} \vDash B$.

We begin our investigation of consequences of this definition by looking at what can be taken over from the statement calculus version. First of all, Theorem 2.4.1 and its corollary carry over unchanged. For example, part (I) of that theorem follows directly from clause (c) of the definition of satisfaction, and the first assertion in (II) follows from clause (e) following the definition of satisfaction. A proof of the corollary may be made by mimicking the earlier one, using Theorem 2.3.3′ in place of Theorem 2.3.3. Thus, as for the statement calculus, the question of what wfs are consequences of others is reduced to that of what wfs are valid. We turn next (as we did earlier for the statement calculus) to a direct method, using rules of inference, for showing that some wf is a consequence of others. Since Theorem 2.4.2 extends to wfs, rules p and t of the statement calculus are available. † Thus $A_1, A_2, \ldots, A_m \vDash B$ if a string $E_1, E_2, \ldots, E_r \ (= B)$ of wfs can be constructed such that the presence of each can be justified by an application of rule p or rule t. For wfs further rules can be supplied. Among them the following two are basic for justifying the presence of a wf E in a string.

Rule (of universal specification) us: E has the form $A(t)$ where t is a term and there is a wf $(x)A(x)$ preceding E such that A admits t for x.

---

* This definition simply makes precise the intuitive idea that "B is true whenever the wfs in X are all true."

† Now, in an application of rule t, validity is to be understood in its present extended sense.

Rule (of universal generalization) ug: E has the form $(x)A$ where A is a preceding wf and x is a variable not having a free occurrence in any premise.

We contend that $A_1, A_2, \ldots, A_m \vDash B$ if we can construct a string $E_1, E_2, \ldots, E_r(=B)$ of wfs such that the presence of each can be accounted for on the basis of one of the rules p, t, us, or ug. Indeed, with this assumption we prove that

$$A_1, A_2, \ldots, A_m \vDash E_i, \ 1 \leq i \leq r.$$

That this is true of E_1 and, if true for $E_1, E_2, \ldots, E_{k-1}$, then true for E_k by virtue of rule p or t, follows from the earlier proof. Suppose next that the assertion holds for $E_1, E_2, \ldots, E_{k-1}$, and E_k is justified by rule us. That is, E_k is $A(t)$ where there is an earlier wf $(x)A(x)$ in the sequence such that A admits t for x. Then $A_1, A_2, \ldots, A_m \vDash (x)A(x)$, and by Theorem 2.8.3(I), $\vDash (x)A(x) \rightarrow A(t)$. Hence $(x)A(x) \vDash A(t)$, by Theorem 2.4.1(I), and so $A_1, A_2, \ldots, A_m \vDash A(t)$ by Theorem 2.4.2(II).

If E_k is justified, instead, by rule ug, then E_k is $(x)A$, where A is a preceding wf and x does not occur free in any premise. Thus $A_1, A_2, \ldots, A_m \vDash A$ and so $\vDash A_1 \wedge A_2 \wedge \cdots \wedge A_m \rightarrow A$ by Theorem 2.4.1(II). From Theorem 2.8.4(I) we deduce that $\vDash A_1 \wedge A_2 \wedge \cdots \wedge A_m \rightarrow (x)A$ and, in turn, that $A_1, A_2, \ldots, A_m \vDash (x)A$. This completes the proof.

In exactly the same way that the rule us is established as a rule of inference using the valid wf $(x)A(x) \rightarrow A(t)$ of Theorem 2.8.3(I), so can any valid wf having the form of conditional be used to derive a rule of inference. Thus, each of the valid wfs having the form of a conditional that are implicit in the wfs of Theorem 2.8.5 yields a rule of inference.

We are now in a position to construct formal derivations of simple arguments in the style developed in §2.4.

Examples

2.9.1. Consider the following argument.

> No human beings are quadrupeds. All women are human beings. Therefore, no women are quadrupeds.

Using the methods of translation of §2.6, we symbolize this as follows.

$$(x)(Hx \rightarrow \sim Qx)$$
$$\frac{(x)(Wx \rightarrow Hx)}{(x)(Wx \rightarrow \sim Qx)}$$

The derivation proceeds as follows.

| {1} | (1) | $(x)(Hx \rightarrow \sim Qx)$ | p |
| {2} | (2) | $(x)(Wx \rightarrow Hx)$ | p |

{2}	(3) $Wy \rightarrow Hy$	2 *us*
{1}	(4) $Hy \rightarrow \sim Qy$	1 *us*
{1,2}	(5) $Wy \rightarrow \sim Qy$	3, 4 *t*
{1,2}	(6) $(x)(Wx \rightarrow \sim Qx)$	5 *ug*

2.9.2. The following argument is more involved.

> Everyone who buys a ticket receives a prize. Therefore, if
> there are no prizes, then nobody buys a ticket.

If Bxy is "x buys y," Tx is "x is a ticket," Px is "x is a prize," and Rxy is "x receives y," then the hypothesis and conclusion may be symbolized as follows.

$$(x)((\exists y)(Bxy \wedge Ty) \rightarrow (\exists y)(Py \wedge Rxy))$$
$$\sim(\exists x)Px \rightarrow (x)(y)(Bxy \rightarrow \sim Ty)$$

Since the conclusion is a conditional, we employ the rule *cp* in the derivation below. The deduction of line 3 from line 2, that of line 7 from line 6, and that of line 11 from 10 should be studied and justified by the reader.

{1}	(1) $(x)((\exists y)(Bxy \wedge Ty) \rightarrow (\exists y)(Py \wedge Rxy))$	*p*
{2}	(2) $\sim(\exists x)Px$	*p*
{2}	(3) $(x) \sim Px$	2 *t*
{2}	(4) $\sim Py$	3 *us*
{2}	(5) $\sim Py \vee \sim Rxy$	4 *t*
{2}	(6) $(y)(\sim Py \vee \sim Rxy)$	5 *ug*
{2}	(7) $\sim(\exists y)(Py \wedge Rxy)$	6 *t*
{1}	(8) $(\exists y)(Bxy \wedge Ty) \rightarrow (\exists y)(Py \wedge Rxy)$	1 *us*
{1,2}	(9) $\sim(\exists y)(Bxy \wedge Ty)$	7, 8 *t*
{1,2}	(10) $(y)(\sim Bxy \vee \sim Ty)$	9 *t*
{1,2}	(11) $(y)(Bxy \rightarrow \sim Ty)$	10 *t*
{1,2}	(12) $(x)(y)(Bxy \rightarrow \sim Ty)$	11 *ug*
{1}	(13) $\sim(\exists x)Px \rightarrow (x)(y)(Bxy \rightarrow \sim Ty)$	2, 12 *cp*

2.9.3. Additional rules of inference may be introduced to expedite derivations and make them more closely resemble the informal proofs that are given in mathematics. We shall discuss two such derived rules. They are formal analogues of two familiar occurrences in everyday mathematics. If one is assured that "$(\exists x)P(x)$" is true, one feels at liberty to "choose" a c such that $P(c)$. Here c is an unknown constant such that $P(c)$. Conversely, given that there is some c such that $P(c)$, one does not hesitate to infer that "$(\exists x)P(x)$" is true. The formal counterparts of these accepted methods of reasoning are the rules to which we wish to draw attention. They may be stated as follows (assuming the preamble that was given prior to the statement of rules *us* and *ug*).

Rule (of existential specification) es: E has the form $A(c)$, where c is a new individual constant and there is a wf $(\exists x)A(x)$ preceding E.

We hasten to add that two restrictions must be imposed on derivations in which this rule is used if one is not to go astray: No application of rule *ug* may be made using a variable which is free in some $(\exists x)B(x)$ to which rule *es* has been previously applied, and the final wf in a derivation may not contain a new individual constant introduced in any application of rule *es*.*

Rule (of existential generalization) eg: E has the form $(\exists x)A(x)$ where $A(x)$ admits t (a term) for x and $A(t)$ precedes E.

The validity of this rule follows immediately from Theorem 2.8.3(II). In the following example illustrating both rules, we employ a lower-case Greek letter to designate an object which is involved in the "act of choice" accompanying an instance of the rule *es*.

> Every member of the committee is wealthy and a Republican. Some committee members are old. Therefore, there are some old Republicans.

{1}	(1)	$(x)(Cx \rightarrow Wx \wedge Rx)$	p
{2}	(2)	$(\exists x)(Cx \wedge Ox)$	p
{2}	(3)	$C\alpha \wedge O\alpha$	2 *es*
{1}	(4)	$C\alpha \rightarrow W\alpha \wedge R\alpha$	1 *us*
{2}	(5)	$C\alpha$	3 *t*
{1,2}	(6)	$W\alpha \wedge R\alpha$	4, 5 *t*
{2}	(7)	$O\alpha$	3 *t*
{1,2}	(8)	$R\alpha$	6 *t*
{1,2}	(9)	$O\alpha \wedge R\alpha$	7, 8 *t*
{1,2}	(10)	$(\exists x)(Ox \wedge Rx)$	9 *eg*

2.9.4. The derivation corresponding to the following argument employs all the rules which we have described.

> Some Republicans like all Democrats. No Republican likes any Socialist. Therefore, no Democrat is a Socialist.

The reason for the introduction of "x" in line 3 below is this. By virtue of the form of the conclusion, $(x)(Dx \rightarrow \sim Sx)$, a conditional proof is given. Thus, Dx is introduced as a premise in line 3. Since x occurs free here, we note its presence (as well as in subsequent lines which depend on this premise) to assist in avoiding any abuse of rule *ug*.

{1}	(1)	$(\exists x)(Rx \wedge (y)(Dy \rightarrow Lxy))$	p
{2}	(2)	$(x)(Rx \rightarrow (y)(Sy \rightarrow \sim Lxy))$	p
{3}	(3)	Dx	x, p
{1}	(4)	$R\alpha \wedge (y)(Dy \rightarrow L\alpha y)$	1 *es*
{1}	(5)	$(y)(Dy \rightarrow L\alpha y)$	4 *t*
{1}	(6)	$Dx \rightarrow L\alpha x$	5 *us*
{1,3}	(7)	$L\alpha x$	x, 3, 6 *t*
{2}	(8)	$R\alpha \rightarrow (y)(Sy \rightarrow \sim L\alpha y)$	2 *us*
{1}	(9)	$R\alpha$	4 *t*
{1,2}	(10)	$(y)(Sy \rightarrow \sim L\alpha y)$	8, 9 *t*

---

* A thorough discussion of rule *es*, including an example to illustrate the need for the first restriction, may be found in Mendelson (1964), pp. 73–75. There it is called "rule C."

{1,2}	(11) $Sx \rightarrow \sim Lax$	10 *us*
{1,2,3}	(12) $\sim Sx$	x, 7, 11 *t*
{1,2}	(13) $Dx \rightarrow \sim Sx$	3, 12 *cp*
{1,2}	(14) $(x)(Dx \rightarrow \sim Sx)$	13 *ug*

The foregoing examples lend plausibility to the contention that the predicate calculus is adequate for formalizing a wide variety of arguments. Lest there be concern over the lengths of derivations of such simple arguments as those considered, we assure the reader that an extended treatment would include the introduction of further derived rules of inference to streamline derivations. The outcome is the concept of an "informal proof." In mathematics this amounts to a derivation in the conversational style to which one is accustomed: mention of rules of inference and tautologies used is suppressed, and attention is drawn only to the mathematical (that is, nonlogical) axioms and earlier theorems employed. (Further details of this are supplied in the next chapter.) The principal advantage which accrues in having informal proofs as the evolution of formal derivations is this: One has a framework within which it can be decided in an objective and mechanical way, in case of disagreement, whether a purported proof is truly a proof.

Exercises

Construct a derivation corresponding to each of the following arguments.

2.9.1. No freshman likes any sophomore. All residents of Dascomb are sophomores. Therefore, no freshman likes any resident of Dascomb. (Fx,Lxy,Sx,Dx)

2.9.2. Art is a boy who does not own a car. Jane likes only boys who own cars. Therefore, Jane does not like Art. (Bx,Ox,Lxy,a,j)

2.9.3. No Republican or Democrat is a Socialist. Norman Thomas was a Socialist. Therefore, he was not a Republican. (Rx,Dx,Sx,t)

2.9.4. Every rational number is a real number. There is a rational number. Therefore, there is a real number.

2.9.5. All rational numbers are real numbers. Some rationals are integers. Therefore, some real numbers are integers. (Qx,Rx,Zx)

2.9.6. All freshmen date all sophomores. No freshman dates any junior. There are freshmen. Therefore, no sophomore is a junior.

2.9.7. No pusher is an addict. Some addicts are people with a record. Therefore, some people with a record are not pushers.

2.9.8. Some freshmen like all sophomores. No freshmen likes any junior. Therefore, no sophomore is a junior. (Fx,Lxy,Sx,Jx)

2.9.9. Some persons admire Elvis. Some persons like no one who admires Elvis. Therefore, some persons are not liked by all persons. (Px,Ex,Lxy)

CHAPTER THREE

Axiomatic Theories

O~NE~ ~OF~ ~THE~ striking aspects of twentieth century mathematical research is the enormously increased role which the axiomatic approach plays. The axiomatic method is certainly not new in mathematics, having been employed by Euclid in his *Elements*. However, only in relatively recent years has it been adopted in parts of mathematics other than geometry. This has become possible because of a fuller understanding of the nature of axioms and the axiomatic method.

In the first three sections of this chapter we discuss the axiomatic method (by way of the notion of an informal axiomatic theory) as it is used currently in everyday mathematics. An understanding of this part of the chapter is adequate preparation for the next chapter. When judged by the standards of current investigations into the foundations of mathematics, our presentation must be classified as antiquated. Indeed, it reflects attitudes about the foundations that prevailed from the 1880's until the 1920's, during which period the axiomatization of various fragments of mathematics was the main activity of workers in the field. An introductory account of the present-day approach to the foundations appears in the remainder of the chapter. Distinctive features of the modern approach include the explicit incorporation among the axioms of a theory those which provide for a "built-in" theory of inference. Thereby, various technical questions pertaining to axiomatic theories can be posed and discussed with precision.

§3.1. Informal Axiomatic Theories. Their Evolution

The concept to be described is an outgrowth of the method used by Euclid in his *Elements* to organize ancient Greek geometry. The plan of this work is as follows. It begins with a list of definitions of such notions as point and line; for example, a line is defined as length without breadth. Next appear various statements, some of which are labeled *axioms* and the others *postulates*. It appears that the axioms are intended to be principles of reasoning which are valid in any science (for example, one axiom asserts that things equal to the same thing are equal to each other), whereas the postulates are intended to be assertions about the subject matter to be discussed—geometry (for example, one postulate asserts that it shall be possible to draw a line joining any two distinct points). From this starting point of definitions, axioms, and postulates, Euclid proceeds to derive propositions (theorems) and at appropriate places to introduce further definitions (for example, an obtuse angle is defined as an angle which is greater than a right angle).

Several comments on Euclid's work are in order. It is clear that his goal was to deduce all the geometry known in his day as logical consequences of certain unproved propositions. On the other hand, we can only conjecture what his attitude was toward other facets of his point of departure. From a modern viewpoint it may be said that he treated point and line essentially as *primitive* or *undefined* notions, subject only to the restrictions stated in the postulates, and that his definitions of these notions offer merely an intuitive description which assists one in thinking about formal properties of points and lines. However, since the geometry of that era was intended to have physical space as an interpretation, it is highly plausible that Euclid assigned physical meaning to these notions. Further evidence to support this conclusion is to be found in some proofs where Euclid made assumptions that cannot be justified on the basis of his primitive notions and postulates, yet which, on the basis of the intended interpretation of his primitive notions, appear to be evident. If, indeed, Euclid confused formal or axiomatic questions and problems concerning applications of geometry, then herein lies the source of the only flaws in his work as judged by modern standards. He probably believed the postulates to be true statements because of the meanings suggested by his definitions of the terms used in them. Since proofs were not provided for the postulates, they acquired the status of "self-evident truths." This attitude toward the nature of postulates or axioms (now, incidentally, no distinction is drawn between these two words) still persists in the minds of many. Indeed, in current nonmathematical writings it is not uncommon to see such phrases as "It is axiomatic that" and "It is a fundamental postulate of" used to mean that some statement is beyond all logical opposition. Within mathematics the attitude toward the nature of axioms has altered radically. The change was gradual and it accompanied the full understanding of the discovery by J. Bolyai and (independently) N. Lobachevsky of a non-Euclidean geometry. Let us elaborate on this matter.

In the traditional sense a non-Euclidean geometry is a geometry whose formulation coincides with that of Euclidean geometry with the one exception that Euclid's fifth postulate (the "parallel postulate") is denied. The fifth postulate is "If two lines are cut by a third so as to make the sum of the two interior angles on one side less than two right angles, then the two lines, if produced, meet on that side on which the interior angle sum is less than two right angles." An equivalent formulation, in the sense that either, together with the remaining postulates, implies the other, and one which is better for making comparisons, is "In a plane, if point A is not on the line l, then there is exactly one line on A parallel to l." This is one of many axioms equivalent to the parallel postulate which were obtained as by-products of unsuccessful attempts to substantiate the belief that the parallel postulate could be derived from Euclid's remaining axioms. Bolyai and Lobachevsky dispelled this belief by developing a geometry in which the parallel postulate was replaced by the statement "In a plane, if the point A is not on line l, then there exists more than one line on A parallel to l." Apparently, the "truth" of this new geometry was initially in doubt. But on the basis of measurements that could be made in the portion of physical space available, there appeared to be no measurable differences between the predictions of the Bolyai-Lobachevsky geometry and those of Euclidean geometry. Also, each geometry, when studied as a deductive system, appeared to be consistent, in not yielding contradictory statements. The ability to examine these geometries from the latter point of view represented a great advance, for, in essence, it amounted to the detachment of physical meaning from the primitive notions of point, line, and so on.

A second advance in the attitude toward the axiomatic method accompanied the creation of various models in Euclidean geometry of the Bolyai-Lobachevsky geometry. A typical example is the model proposed in 1871 by Felix Klein, for which he interpreted the primitive notions of plane, point, and line, respectively, as the interior of a fixed circle in the Euclidean plane, a Euclidean point inside this circle, and an open-ended chord of this circle. If, in addition, distances and angles are computed by formulas developed by A. Cayley, in 1859, then all axioms of plane Bolyai-Lobachevsky geometry become true statements. The immediate value of such an interpretation was to establish the relative consistency (a concept which will be described in detail later) of the Bolyai-Lobachevsky geometry. That is, if Euclidean geometry is a consistent logical structure, then so is the Bolyai-Lobachevsky geometry. Of greater significance, for understanding the nature of axiomatic theories, was the entertainment of the possibility of varying the meaning of the primitive notions of an axiomatic theory while holding fixed its deductive structure.

This evolution in the understanding of the axiomatic method set the stage for the present-day concept of an informal axiomatic theory. Such a theory is presented in essentially the same way that Euclid began his development of geometry—by listing the primitive notions and the axioms of the theory. However, the *attitude* about these has changed: *Now* the primitive notions are taken to be undefined and the axioms are regarded as assumptions about them. Specifically, each axiom states a property that is assumed about an undefined term or a relation that is assumed to hold among two or more such terms. Thus, collectively, the axioms constitute an initial stock of theorems of the theory. The question of whether an axiom is *true* becomes relevant only when a meaning is assigned to the pertinent undefined terms. In principle, primitive notions and axioms for them may be concocted in an arbitrary manner. In practice this is not done, for the whole point of an axiomatic theory is to organize some subject matter in a systematic way. Thus, one who formulates an axiomatic theory has in mind at least *one* interpretation of the primitive terms for which the axioms become true statements. We shall elaborate on these matters as we discuss the development of informal axiomatic theories from intuitive theories.

Usually one's first exposure to some branch of science is by way of an intuitive approach; subjects such as arithmetic, geometry, mechanics, and set theory, to cite just a few, are approached in this way. An axiomatization of such an intuitive theory can be attempted when the fundamental notions and properties are believed to be known and the theory appears to be sound enough that reliable predictions can be made with it. The first step in such an attempt is to list what are judged to be the basic notions discussed by the theory, together with what are judged to be a basic set of true statements about these notions. In order to carry out this step efficiently, one often elects to presuppose certain theories previously constructed. In informal axiomatics within mathematics, one usually assumes a theory of logic along with a theory of sets. In axiomatic work in an empirical science such as economics or physics it is standard procedure to assume, in addition to logic and set theory, parts of classical mathematics. Once it has been decided what theories will be assumed, the key steps in the axiomatization can be carried out. The first of these is the introduction of symbols (including, possibly, words) as names for those notions which have been judged to be basic for the intuitive theory. These are called the **primitive symbols** (or **terms**) of the axiomatic theory. The only further symbols which are admitted (aside from symbols of the presupposed theories) are **defined symbols,** that is, expressions whose meaning is explicitly stated in terms of the primitive symbols. (The intuitive theory in mind often suggests the introduction of some such symbols.) The next step is the translation of those statements that were singled out as expressing fundamental properties of the basic notions of the intuitive theory into the language which can be constructed from just the primitive and defined terms (and those of any theory which is presupposed). Illustrations of such languages appear in the Examples in §3.7.

In a program of the sort we have described for axiomatizing an intuitive theory, there is often considerable leeway in the choice of primitive notions. Different sets may be suggested by various combinations of notions which occur in the intuitive theory. In the modern axiomatization of Euclidean geometry devised by D. Hilbert there are six primitive notions: point, line, plane, incidence, betweenness, and congruence. On the other hand, in that created by M. Pieri there are but two primitive notions: point and motion. Obviously the choice of primitive notions for an axiomatic theory influences the choice of axioms. A great variety of more subtle remarks can be made concerning the selection of axioms for a particular theory. Some are presented in §3.4.

While we are dealing in generalities we will mention another stimulus for the creation of axiomatic theories—the observation of basic likenesses in the central features of several different theories. This may prompt an investigator to distill out these common features and use them as a guide for defining an axiomatic theory in the manner described above. Any one of the theories which an axiomatic theory is intended to formalize serves as a potential source of definitions and possible theorems of this axiomatic theory. An axiomatic theory which successfully formalizes an intuitive theory is a source of insight into the nature of that theory, since the axiomatic theory is developed without reference to meaning. One which formalizes each of several theories to some degree has the additional merit that it effects simplicity and efficiency. Since such an axiomatic theory has an interpretation in each of its parent theories (on a suitable assignment of meaning to its primitive terms), it produces simplicity because it tends to reduce the number of assumptions which have to be taken into account for particular theorems in any one of the parent theories. Efficiency is effected, because a theorem of the axiomatic theory yields a theorem of each of the parent theories. Herein lies one of the principal virtues of taking the primitive terms of an axiomatic theory as undefined.

A by-product of the creation of an axiomatic theory which is the common denominator of several theories is the possibility of enriching and extending given theories in an inexpensive way. For example, a theorem in one theory may be the origin of a theorem in the derived theory and it, in turn, may yield a new result in another parent theory. In addition to the possible enrichment in content of one theory by another, by way of an axiomatic theory derived from both, there is also the possibility of "cross-fertilization" of methods of attack on problems. That is, a method of proof which is standard for one theory may provide a new method in another theory, with a derived theory serving as the linkage.

A full understanding of such remarks as the foregoing cannot possibly be achieved until one has acquired some familiarity with a variety of specific theories and analyzed some successful attempts to bring diverse theories under a single heading. The field of algebra abounds in such successful undertakings. Indeed, it is perhaps in algebra that this type of genesis and exploitation of theories has scored its greatest successes.

§3.2. Informal Axiomatic Theories. Examples

We have said that informal axiomatic theories presuppose a theory of inference and a theory of sets. The working forms which are adopted for these two assumed theories should be thoroughly understood; so we shall give this matter our attention. The theory of inference in question is simply the intuitive theory which one absorbs by studying mathematics! That this theory is clearly defined is suggested by the fact that what is judged to be a proof by one competent mathematician is usually acceptable to other mathematicians. More substantive remarks can be offered. In §2.9 appeared fragments of the theory of inference supplied by first-order logic. That the rules of inference given there formalize some aspect of intuitive reasoning is substantiated by the examples that were discussed. The development of rules of inference could be continued to the point where there would be considerable evidence to support the contention that the definition of logical correctness supplied by first-order logic is closely attuned to the corresponding intuitive notion which mathematicians acquire. Such a book as *Logic for Mathematicians*, by J. B. Rosser (1953), is rich in examples which illustrate his thesis that logical principles which are judged as correct by most mathematicians are classified as correct by symbolic logic and vice versa. That is, there is considerable evidence in support of the thesis that the theory of inference which is presupposed for an informal theory is clearly defined and can be spelled out if necessary.*

The theory of sets which is assumed for an informal theory is that which stems from the groundwork laid in Chapter 1. Although contradictions can be devised within this intuitive theory, that part which is employed in developing informal theories does not lead to such difficulties insofar as is known. We support this latter statement with only the following remark. The intuitive set theory based on the principles set forth in Chapter 1 can be axiomatized in such a way that (i) so far as is known, all undesirable features are avoided (that is, the derivations which yield such contradictions as the Russell paradox in naive set theory do not translate into derivations of contradictions in the axiomatized version) and (ii) all desirable features consonant with (i) are retained.†

We turn now to some examples of informal theories. These will illustrate the two circumstances described at the end of the preceding section under which axiomatic theories are devised (namely, to formalize simultaneously several theories and to axiomatize some one intuitive theory).

Examples

3.2.1. Our first example of an informal theory is based on likenesses which one might observe in two familiar mathematical systems. The first system consists of the set $G(X)$ of all one-to-one mappings on a nonempty set X onto itself,

---

* The "spelling out" which we have suggested is not without its difficulties. As was pointed out in §2.9, the question of whether some wf is a consequence of others is equivalent to that of whether a certain wf is valid. But, as mentioned in §2.8, employing a plausible definition of "recipe," it can be proved that there is no recipe to test for validity. This state of affairs is why first-order logic is itself studied from an axiomatic point of view (see §3.6).

† Furthermore, although in the customary systems of axiomatic set theory it is impossible to derive the known antinomies of intuitive set theory, for none of these systems has freedom from all contradictions been shown (at least by means commonly agreed upon by logicians from various schools).

together with function composition. The second system consists of the set $\mathbb{Z}$ of integers together with the familiar operation of addition. On the basis of results obtained in Chapter 1, and our knowledge of the number system, we may assert that: (i) each operation is a binary operation in the corresponding set; (ii) each operation is associative; (iii) in each set there is a distinguished element [namely, i_X in $G(X)$ and 0 in $\mathbb{Z}$] such that when it is combined with any element of the set to which it belongs the same element results; and (iv) for each element of either set there exists an element of the same set such that when the two are combined the distinguished element of that set results [that is, $f \circ f^{-1} = f^{-1} \circ f = i_X$ and $x + (-x) = (-x) + x = 0$].

The similarities in these two systems are incorporated in the informal axiomatic theory called **group theory**. The primitive terms are an unspecified set G, a binary operation in G, for which we use multiplicative notation (that is, the operation will be symbolized by $\cdot$ and the value at $\langle a,b \rangle$ of this function on $G \times G$ into G will be designated by $a \cdot b$), and an element e of G. The axioms are the following.

> G_1. For all a, b, and c in G, $a \cdot (b \cdot c) = (a \cdot b) \cdot c$.
>
> G_2. For all a in G, $a \cdot e = e \cdot a = a$.
>
> G_3. For each a in G there exists an a' in G such that $a \cdot a' = a' \cdot a = e$.

The above is a formulation of group theory as one might find it in an algebra text. In harmony with the agreement to write the value of $\cdot$ at $\langle a,b \rangle$ as $a \cdot b$, we call this element the *product* of a and b. Henceforth we shall use the simpler notation ab for it. An element which has the property assumed for e in G_2 is called an *identity element* and an element which satisfies G_3 for a given a is called an *inverse* of a (relative to e).

A few theorems of group theory are proved next.

> G_4. G contains exactly one identity element.

Proof. In view of G_2, only a proof of the uniqueness is required. Assume that each of e_1 and e_2 is an identity element of G. Then $e_1 a = a$ for every a, and $ae_2 = a$ for every a. In particular, $e_1 e_2 = e_2$ and $e_1 e_2 = e_1$. Hence, $e_1 = e_2$ by properties of equality.

> G_5. Each element in G has exactly one inverse.

Proof. Since G_3 asserts the existence of an inverse for each element a, only the proof of its uniqueness remains. Assume that both a' and a'' are inverses of a. Then $a''a = e$ and $aa' = e$. By G_1, $(a''a)a' = a''(aa')$, and, hence, $ea' = a''e$. Using G_2 it follows that $a' = a''$.

In multiplicative notation the inverse of a is designated by "a^{-1}"; thus $a^{-1}a = aa^{-1} = e$ (the unique identity element of G).

> G_6. For every a, b, and c in G, if $ab = ac$, then $b = c$, and, if $ba = ca$, then $b = c$.

Proof. Assume that $ab = ac$. Now $a^{-1}(ab) = (a^{-1}a)b = eb = b$. On the other hand, $a^{-1}(ab) = a^{-1}(ac) = (a^{-1}a)c = ec = c$. Hence, $b = c$. The proof of the remaining assertion is similar.

Proofs of the next two theorems are left as exercises.

G_7. For all a and b in G, each of the equations $ax = b$ and $ya = b$ has a unique solution in G.

G_8. For all a and b in G, $(ab)^{-1} = b^{-1}a^{-1}$.

3.2.2. The theory to be described has its origin in Euclidean plane geometry. It is that generalization of Euclidean geometry known as **affine geometry.** The primitive terms are a set $\mathcal{P}$ (whose members are called *points* and will be denoted by such letters as P, Q, . . .), a set $\mathcal{L}$ (whose members are called *lines* and will be denoted by such letters as l, m, . . .), and a set $\mathcal{J}$ called the *incidence relation.* The axioms are as follows.

AG_1. $\mathcal{J} \subseteq \mathcal{P} \times \mathcal{L}$. ($\langle P, l \rangle \in \mathcal{J}$ is read "P lies on l," or "l contains P," or "l passes through P").

AG_2. For any two distinct points P and Q there is exactly one line passing through P and Q. (This line will be denoted by $P + Q$.)

AG_3. For any point P and any line l there exists exactly one line m passing through P and *parallel* to l (that is, either $m = l$ or there exists no point which lies on both l and m).

AG_4. If A, B, C, D, E, and F are six distinct points such that $A + B$ is parallel to $C + D$, $C + D$ is parallel to $E + F$, $A + C$ is parallel to $B + D$, and $C + E$ is parallel to $D + F$, then $A + E$ is parallel to $B + F$.

AG_5. There exist three distinct points not on one line.

Proofs of a few simple theorems are called for in the exercises.

3.2.3. In §1.12 appeared a formulation of the theory of natural numbers as an informal axiomatic theory. The primitive terms are a set D, a unary operation s in D, and an element 0 of D. The axioms are the assumptions P1 to P3 stated there. The formulation which Peano himself gave is a slight variation of the foregoing. In it the primitive terms are *natural number*, *zero* (0), and *successor* $(')$, and the axioms are as follows.

1. 0 is a natural number.
2. If n is a natural number, then n' is a natural number.
3. For any natural numbers m and n, if $m' = n'$, then $m = n$.
4. For any natural number n, $n' \neq 0$.
5. If M is a set of natural numbers such that (i) $0 \in M$ and (ii) whenever $m \in M$ then $m' \in M$, then M is the set of all natural numbers.

Our examples illustrate the fact that a variety of mathematical theories are formulated in terms of a nonempty set D, a set of relations in D,

a set of operations in D, and a (possibly empty) set of elements of D,
Although either the set of relations or the set of functions may be empty,
their union is nonempty. Collectively, these relations, functions, and
elements of D serve as the basis for imposing a certain structure on D,
The structure itself is given in the axioms, which are those properties
assigned to D and the associated relations, functions, and elements of D
(including, possibly, the presence of inner relations among them). To
systematize the formulation of theories that fit this pattern we introduce
the following notion.

A (**mathematical**) **structure** is an ordered quadruple

$$\mathfrak{A} = \langle D, \langle R_i \rangle_{i<k}, \langle F_i \rangle_{i<l}, \langle d_i \rangle_{i<m} \rangle$$

where D is a nonempty set (called the domain of the structure), k, l, and
m are specified natural numbers, R_i is an m_i-ary (m_i specified) relation
in D, F_i is an n_i-ary (n_i specified) operation in D, and $d_i \in D$. The constants which enter into the definition of $\mathfrak{A}$ will be written as the ordered
quintuple

$$\sigma = \langle k,l,m, \langle m_i \rangle_{i<k}, \langle n_i \rangle_{i<l} \rangle.$$

This is called the **signature** of $\mathfrak{A}$. If k, l, or m is 0, it is understood that
the corresponding set is absent. In discussing examples it is convenient
to use the variant

$$\mathfrak{A} = \langle D, R_0, R_1, \ldots, F_0, F_1, \ldots, d_0, d_1, \ldots \rangle$$

of the above notation for a structure and state the values of m_i and n_i for
each i in the running text.

The approach to the axiomatization of theories that is possible using
the concept of a structure is *defining* axiomatic theories by set-theoretical
predicates. The consideration of examples will bring the procedure into
focus. In the first we consider the theory of partially ordered sets. The
purely set-theoretical character of the predicate "is a partially ordered
set" which is defined should be apparent.

DEFINITION A. $\mathfrak{A}$ is a partially ordered set iff it is a structure
$\langle X,\rho \rangle$* where ρ is a binary relation in X and

0_1. ρ is reflexive on X,
0_2. ρ is antisymmetric,
0_3. ρ is transitive.

---

* In such a definition "$\langle X,\rho \rangle$" is used ambiguously. The notation together with the phrase
"where ρ is a binary relation in X" serves to designate any structure having signature
$\langle 1,0,0,\langle 2 \rangle \rangle$. (Here the fifth coordinate is missing, since no functions are present.)

The sentence which makes up the definition may need recasting if it is to appear in the running text, since it begins with a symbol. The following version meets this objection.

DEFINITION A'. A partially ordered set is a structure $\langle X, \rho \rangle$ where ρ is a binary relation in X such that

O_1. ρ is reflexive on X,
O_2. ρ is antisymmetric,
O_3. ρ is transitive.

An alternative to Definition A which is closer to standard mathematical practice is a conditional definition.

DEFINITION A''. Let X be a set and ρ be a binary relation in X. Then the structure $\langle X, \rho \rangle$ is a partially ordered set iff

O_1. ρ is reflexive on X,
O_2. ρ is antisymmetric,
O_3. ρ is transitive.

This definition is conditional in the sense that the proper definition is preceded by a hypothesis. When a definition is so formulated, it is common practice to omit the hypothesis in stating theorems of the theory.

Our second example is a definition of group theory along the lines suggested by the axiomatization appearing in Example 3.2.1.

DEFINITION B. $\mathfrak{G}$ is a group iff it is a structure $\langle G, \cdot, e \rangle$ where $\cdot$ is a binary operation in G, e is an element of G, and

G_1.   for all a, b, and c in G, $a \cdot (b \cdot c) = (a \cdot b) \cdot c$,
G_2.   for all a in G, $a \cdot e = e \cdot a = a$,
G_3.   for each a in G there is an a' in G such that $a \cdot a' = a' \cdot a = e$.

The definition of affine geometry in Example 3.2.2 may appear to be outside the class of theories that can be defined in terms of a structure because two basic sets are included among the primitive terms. This difficulty is surmountable—one considers a structure $\mathfrak{A}$ whose domain, D, is thought of as the union of the set of points and the set of lines. Then two unary relations, whose respective roles are to identify points and lines, are assigned to $\mathfrak{A}$. The remaining coordinate of $\mathfrak{A}$ is a binary relation in D; its intended interpretation is the incidence relation.

When a theory $\mathfrak{T}$ is formulated in terms of a set-theoretical predicate involving a structure $\langle D, R_0, \ldots \rangle$, the coordinates of this n-tuple are symbols for the primitive terms of $\mathfrak{T}$. As used in the formulation of $\mathfrak{T}$, these symbols denote sets which are unspecified apart from meeting certain requirements (for example, that R_0 be a binary relation in D). If, upon the assignment of a value to each (consistent with the given requirements), the resulting "concrete" structure $\mathfrak{A}$ satisfies the predicate (thus, satisfies the axioms which are present), then $\mathfrak{A}$ is called a **model** of $\mathfrak{T}$.

Remark.

This definition is not at odds with that of a "model" as defined in §2.8. To substantiate this assertion we proceed as follows. Consider the class of structures having $\sigma = \langle k,l,m,\langle m_i \rangle_{i<k},\langle n_i \rangle_{i<l} \rangle$ as signature. Let us form the first-order language, call it L_σ, having $P_i^{m_i}$, $i < k$, as predicate letters, $f_i^{n_i}$, $i < l$, as function letters, and a_i, $i < m$, as individual constants. This language is designated to "talk about" each structure of the class having signature σ and, hence, is the first-order language accompanying a theory $\mathfrak{T}$ defined in terms of σ and certain axioms. Now consider the interpretation $I = \langle D,g \rangle$ of L_σ such that $g(P_i^{m_i}) = R_i$, $i < k$, $g(f_i^{n_i}) = F_i$, $i < l$, and $g(a_i) = d_i$, $i < m$. Suppose that I is a model, in the sense of §2.8, of a set X of wfs of L_σ, that is, each A in X is true for I. This means that, intuitively, the translation of each A (as specified by I) is true. Now I determines a structure, namely $\mathfrak{A} = \langle D,\langle R_i \rangle_{i<k},\langle F_i \rangle_{i<l},\langle d_i \rangle_{i<m} \rangle$, and clearly the translation of each A is a true statement about $\mathfrak{A}$. Thus, if I is a model of the set of axioms each axiom is a true statement about $\mathfrak{A}$. So, the definition in §2.8 is the formal analogue of that given above, for according to the latter, if $\mathfrak{A}$ is a model of $\mathfrak{T}$, then the axioms of $\mathfrak{T}$ are true statements about $\mathfrak{A}$.

Examples

3.2.4. With group theory formulated as in Definition B above, let us interpret G as the set $G(X)$ of all one-to-one correspondences of a nonempty set X with itself, interpret the binary operation assigned to G as function composition in $G(X)$, and take the distinguished element e of G as i_X, the identity map on X. The axioms G_1 to G_3 are true for this interpretation, and hence the system consisting of $G(X)$, function composition in $G(X)$, and i_X is a model of group theory. One usually says simply that $G(X)$, together with function composition and i_X, is a group. Next, interpret G as $\mathbb{Z}$, the binary operation in G as ordinary addition in $\mathbb{Z}$, and e as the integer 0. The axioms are true for this interpretation by virtue of familiar properties of addition, and so this system is a group. Again, $\mathbb{R}^+$ together with ordinary multiplication and 1 is found to be a group. The same is true of the power set of a set together with the operation of symmetric difference and $\varnothing$.

3.2.5. Referring to Example 3.2.2, one who is familiar, to some degree, with Euclidean geometry will undoubtedly accept it as an affine geometry. A radically different model results on setting $\mathcal{P} = \{1,2,3,4\}$, $\mathcal{L} = \{\{1,2\}, \{1,3\}, \{1,4\}, \{2,3\}, \{2,4\}, \{3,4\}\}$, and defining P to lie on l iff $P \in l$. The verification that all axioms are satisfied is left as an exercise.

3.2.6. A model of the theory of natural numbers (see Example 3.2.3) results upon choosing D to be the sequence $\{a, ar, \ldots, ar^n, \ldots\}$ with a and r nonzero, and s to be the function such that $s(ar^n) = ar^{n+1}$.

It is an accepted property of a model $\mathfrak{M}$ of an informal theory $\mathfrak{T}$ that each theorem of $\mathfrak{T}$ is true in $\mathfrak{M}$. The supporting argument is simply that (by definition of a model of $\mathfrak{T}$) each axiom is true in $\mathfrak{M}$ and each theorem of $\mathfrak{T}$ is derived from the axioms by logic alone. An illustration may be given in terms of Theorem G_8 of Example 3.2.1. The interpretation of G_8 in the group $G(X)$ of mappings is the statement that if $a, b \in G(X)$, then $(a \circ b)^{-1} = b^{-1} \circ a^{-1}$, which is an important property of functional inversion. The interpretation of G_8 in the group consisting of $\mathbb{Z}$, addition, and 0 is the statement that $-(a+b) = (-b) + (-a)$. Thus, these two results, diverse in appearance, are interpretations of a single statement of group theory.

Exercises

3.2.1. Prove Theorems G_7 and G_8 in Example 3.2.1.

3.2.2. The theory of commutative groups differs from the theory of groups in that it includes one further axiom:

G_9. For all a and b in G, $ab = ba$.

It is common practice to use additive notation for the operation in a commutative group (that is, to write $a + b$ instead of ab), to write 0 instead of e, and to write $-a$ instead of a^{-1}.

Suppose that G together with $+$ and 0 is a commutative group. Prove each of the following theorems.

(a) $-(a + b) = (-a) + (-b)$.

(b) If "$a - b$" is an abbreviation for "$a + (-b)$," then $a + b = c$ iff $b = c - a$.

(c) $a - (-b) = a + b$ and $-(a - b) = b - a$.

(d) If $f \colon G \longrightarrow G$ where $f(a) = -a$, then f is a one-to-one and onto mapping.

3.2.3. Let $\mathbb{Z}_n$ be the set of residue classes $[a]$ of $\mathbb{Z}$ modulo n (see Section 1.7). Show that the relation $\{\langle\langle[a], [b]\rangle, [a + b]\rangle | [a], [b] \in \mathbb{Z}_n\}$ is a binary operation in $\mathbb{Z}_n$. Show that $\mathbb{Z}_n$ together with this operation and $[0]$ is a commutative group.

3.2.4. Show that an operation $+$ can be introduced in the set I of equivalence classes defined in Exercise 1.7.11 by the definition $[a,b] + [c,d] = [a + c, b + d]$, where $[a,b]$ is the equivalence class determined by $\langle a,b \rangle$, and so on. Prove that I together with this operation and $[1,1]$ is a commutative group.

3.2.5. Show that $\mathbb{R}$ together with the operation $\star$ such that $x \star y = (x^3 + y^3)^{1/3}$ and 0 is a group.

3.2.6. Write out the elements of $G(X)$ for $X = \{1,2\}$ and for $X = \{1,2,3\}$. Show that the group associated with the latter set of mappings is not commutative.

3.2.7. Suppose that G is a nonempty set, that $\cdot$ is a binary operation in G, and that G_1 and G_7 hold. Prove that G with $\cdot$ is a group.

3.2.8. Suppose that G is a nonempty finite set, that $\cdot$ is a binary operation in G, and that G_1 and G_6 hold. Prove that G with $\cdot$ is a group.

3.2.9. This exercise is concerned with affine geometry as formulated in Example 3.2.2.

(a) Prove that "is parallel to" is an equivalence relation on L. An equivalence class is called a *pencil* of lines.

(b) Let L_1 and L_2 be two pencils of lines. Using only AG$_2$ and AG$_3$, establish a one-to-one correspondence between the points contained in a line l of L_1 and the lines of L_2.

(c) Using (b), deduce that if there exist three distinct pencils of parallel lines, then there exists a one-to-one correspondence between the lines of any two pencils.

(d) From AG$_5$ infer that there exist at least three distinct pencils of lines.

(e) Show that the set of four points and six lines given in the text is a model of the theory.

(f) Show that any affine geometry contains at least four points and six lines.

3.2.10. Let $\mathfrak{S}$ be the axiomatic theory having as its primitive terms two sets P and L and as its axioms the following.

A$_1$. If $l \in L$, then $l \subseteq P$.

A$_2$. If a and b are distinct elements of P, then there exists exactly one member l of L such that $a, b \in l$.

A$_3$. For every l in L there is exactly one l' in L such that l and l' are disjoint.

A$_4$. L is nonempty.

A$_5$. Every member of L is finite and nonempty.

Establish the following theorems for $\mathfrak{S}$.

(a) Each member of L contains at least two elements.

(b) P contains at least four elements.

(c) L contains at least six elements.

(d) Each member of L contains exactly two elements.

The following exercises are concerned with the theory of simply ordered commutative groups, which may be defined as follows: $\mathfrak{G}$ is a simply ordered commutative group (*s.o.c.g.*) iff $\mathfrak{G} = \langle G, \leq, +, 0 \rangle$, where

SG$_1$. $\langle G, +, 0 \rangle$ is a commutative group;

SG$_2$. $\langle G, \leq \rangle$ is a simply ordered set;

SG$_3$. for all a, b, and c in G, if $a < b$, then $a + c < b + c$. (Here, "$a < b$" is an abbreviation for "$a \leq b$ and $a \neq b$.")

All results obtained earlier for groups, in particular, commutative groups, may be used when needed. Also, properties of simply ordered sets may be used.

3.2.11. Find two *s.o.c.g.* within the real number system.

3.2.12. If $\langle G, \leq, +, 0 \rangle$ is a *s.o.c.g.*, define G^+ to be $\{a \in G | 0 < a\}$. Prove the following properties of G^+.

(a) If $a \in G^+$, then $-a \notin G^+$.

(b) If $a \neq 0$, then either $a \in G^+$ or $-a \in G^+$.

(c) If $a, b \in G^+$, then $a + b \in G^+$.

3.2.13. Prove the following theorems for a *s.o.c.g.*

(a) If $a < b$, then $a - c < b - c$.

(b) If $a + c < b + c$, then $a < b$.

(c) If $a < b$ and $c < d$, then $a + c < b + d$.

(d) If $a < b$, then $-b < -a$.

3.2.14. Prove the following theorem. If G has more than one element and $\langle G, \leq, +, 0 \rangle$ is a *s.o.c.g.*, then G has infinitely many elements.

§3.3. Informal Axiomatic Theories. Further Features

In this section we introduce a variety of notions relevant to informal theories. Most of these provide a classification scheme for a given theory by which its status and its merits can be summarized concisely.

Suppose that A is a formula of some theory $\mathfrak{T}$ and that both A and $\sim A$ are theorems. Then, if the system of logic employed includes the statement calculus with modus ponens as a rule of inference, any formula B of the theory is a theorem. Indeed, $A \rightarrow (\sim A \rightarrow B)$ is a theorem since it is a tautology, and two uses of modus ponens establish B as a theorem. A theory $\mathfrak{T}$ is called **inconsistent** if it contains a formula A such that both A and $\sim A$ are theorems. A theory is called **consistent** if it is not inconsistent—that is, if it contains no formula A such that both A and $\sim A$ are theorems.

Since in any theory we consider the logical apparatus will include what was used above, we regard an inconsistent theory as worthless, since every formula is a theorem. Thus, the question of establishing the consistency of a theory becomes of primary importance. A moment's reflection will point out the high degree of improbability of reaching an answer by direct application of the definition and, consequently, of the need for a "working form" of the definition of consistency. That which is usually adopted in mathematics is: the existence of a model of a theory implies the consistency of the theory. The supporting argument is based on (i) the property of a model mentioned at the end of §3.2, namely, if $\mathfrak{M}$ is a model of the theory $\mathfrak{T}$, then each theorem of $\mathfrak{T}$ is true in $\mathfrak{M}$, and (ii) the assumption that if S is a $\mathfrak{T}$-statement,* then not both of S and $\sim S$ are true in $\mathfrak{M}$. Indeed, assuming (i) and (ii), suppose that $\mathfrak{T}$ has a model $\mathfrak{M}$. If both of the $\mathfrak{T}$-statements S and $\sim S$ are theorems, then both S and $\sim S$ are true in $\mathfrak{M}$ by (i) and this is a contradiction by (ii). Hence, if $\mathfrak{T}$ has a model, then $\mathfrak{T}$ is consistent.

In essence, the foregoing working form of consistency merely substitutes an inspection of true statements about a model of a theory for an inspection of theorems of the theory. If a model of a theory defined in terms of a structure having D as its domain can be found such that the interpretation of D is a *finite* set, one may expect that the question of whether it is free from contradiction can be settled by direct observation. For example, the fact that $\langle \{e\}, \, \cdot, \, e \rangle$, where $e \cdot e = e$, is a model of group theory establishes the consistency of group theory beyond all doubt.

If, on the other hand, a theory has only infinite models (that is, models where the interpretations of the basic set are infinite), then no net gain results upon substituting an inspection of true statements about a model for that of theorems of the theory. Such models of a given theory $\mathfrak{T}$ really amount to interpretations of $\mathfrak{T}$ in another theory such that the interpretation of each axiom of $\mathfrak{T}$ is a *theorem* of the other theory. If this other theory is consistent, then $\mathfrak{T}$ must be. For suppose that a contradiction were deducible from the axioms of $\mathfrak{T}$. Then, in the other theory, by corresponding inferences about the objects constituting the model, a contradiction would be deducible from the corresponding theorems. Such demonstrations of consistency are merely relative: The theory for which a model is devised is consistent if that from which the model is taken is consistent. Let us consider some examples. As described in §3.1, the plane geometry of Bolyai-Lobachevsky has a model in Euclidean plane geometry. Thereby the relative consistency of this non-Euclidean geometry is established in the form: If Euclidean geometry is consistent, then so is the Bolyai-Lobachevsky geometry. A proof of

---

* That is, S is a closed wf of the first-order language in which $\mathfrak{T}$ can be formulated.

the consistency of Euclidean geometry, as precisely formulated in Hilbert (1899), can be given by interpreting a point as an ordered pair of real numbers and a line as a linear equation; in more familiar guise this is simply the standard coordinatization of the Euclidean plane. However, since the theory of real numbers has never been proved consistent, one may conclude merely that *if* the theory of real numbers is consistent, then so is Euclidean geometry. In other words, we obtain a *relative* consistency proof. In turn, since a construction of the real numbers can be given, starting from Peano's axioms, within a sufficiently rich theory of sets, a consistency proof of the theory of real numbers can be given relative to a theory which embraces both Peano's theory and this theory of sets.

Assuming that the consistency of a theory has been settled in the affirmative by proof or by faith, the question of its completeness may be raised. In rough terms, a theory is called complete if it has enough theorems for some purpose. The variety of purposes which may enter in this connection are responsible for a variety of technical meanings being assigned to this notion. However, most definitions of completeness fit into either the category which corresponds to a positive approach or that which corresponds to a negative approach to the question of a sufficiency of theorems. We shall give one definition in the first category and two in the second. The setting for the first of these, which is in the positive vein, is as follows. We know that if $\mathfrak{M}$ is a model of a theory $\mathfrak{T}$ and T is a theorem of $\mathfrak{T}$, then T is true in $\mathfrak{M}$. We might regard $\mathfrak{T}$ as being complete *with respect to* $\mathfrak{M}$ if, conversely, whenever a $\mathfrak{T}$-statement has a true statement of $\mathfrak{M}$ as its interpretation, then that $\mathfrak{T}$-statement is a theorem. This suggests calling $\mathfrak{T}$ complete if it is complete with respect to every model. If we understand by a (universally) valid statement of a theory one which is true in every model, then the notion of completeness which we have in mind may be formulated as: A theory $\mathfrak{T}$ is **deductively complete** iff every valid statement of $\mathfrak{T}$ is provable. The statement calculus can be formulated as an axiomatic theory which is complete in this sense (see §3.5); that is, every tautology is a theorem.

If we approach the question of a sufficiency of theorems in a negative fashion, we are led to a second category of formulations of completeness. For example, we might say that a theory is complete if the axioms provide all theorems we can afford to have without some dire consequence (such as inconsistency) ensuing. A circumstance which might suggest this interpretation of completeness is an attempt to devise an axiomatic theory intended to formalize some intuitive theory. For then one strives to include enough axioms that as many as possible true propositions of the intended model can be obtained as interpretations of theorems of the theory. Hence, one keeps adding, as axioms, formulas which express true propositions of the model up to the point that an inconsistent theory results. This approach to completeness may be crystallized in the following definition. An axiomatic theory $\mathfrak{T}$ is **formally complete** provided that any theory $\mathfrak{T}'$, which results from $\mathfrak{T}$ by the adjunction to the axioms of $\mathfrak{T}$ of a statement of $\mathfrak{T}$ which is not already a theorem of $\mathfrak{T}$, is inconsistent. A consistent theory which is formally complete may be said to have maximum consistency.

An axiomatic theory $\mathfrak{T}$ is said to be **negation complete** if, for any statement A of the theory, either A or $\sim A$ is a theorem. It is clear that negation completeness implies formal completeness. Conversely, if the theory of inference employed in developing an axiomatic theory includes a deduction theorem—that is, a theorem which asserts that if a formula B is deducible from formulas $A_1, A_2, \ldots, A_m$, then $A_m \to B$ is deducible from $A_1, A_2, \ldots, A_{m-1}$—then formal completeness implies negation completeness. To show this, suppose that a theory $\mathfrak{T}$ is formally complete and that the $\mathfrak{T}$-statement A is not a theorem. Then the theory which results on the adjunction of A as an axiom is inconsistent. That is, if Γ is the set of axioms of $\mathfrak{T}$, then a contradiction C can be derived from $\Gamma \cup \{A\}$, whence $A \to C$ can be derived from Γ. In turn, since $(A \to C) \to \sim A$ is a theorem (being a tautology), $\sim A$ can be derived from $A \to C$. Hence, $\sim A$ can be derived from Γ; that is $\mathfrak{T}$ is negation complete.

We may loosely relate consistency and completeness in the following way. An axiomatic theory is consistent if it does not have too many theorems and it is complete if it does not have too few. If an axiomatic theory is both consistent and negation complete, then all questions which arise within the framework of the theory are theoretically decidable in exactly one way. For any statement of the theory is either provable or **refutable** (that is, its negation is provable) because of completeness, and cannot be both proved and refuted because of consistency. Such a state of affairs for a theory does not always imply that proofs or refutations of specific statements of the theory are automatically made available, but in some interesting cases it does. That is, for *some* consistent and complete theories there exists a method which can be described in advance for deciding in a finite number of steps whether a given formula of the theory is a theorem. Such theories are called decidable (see §3.7).

Notions of the sort which we have introduced so far in this section as well as that of categoricity (which is described next) cannot, in general, be discussed in a precise and definitive way at our present intuitive level of discourse. A precise account is possible only when the theory of inference is explicitly incorporated into an axiomatic theory.

The remaining notion which we shall introduce as an ingredient of a classification scheme arises in connection with the purpose for which a theory is devised. An axiomatic theory designed to formalize an intuitive development of a particular mathematical system, for example, Euclidean plane geometry or the natural number system, might be judged as capturing all basic properties of the system only if it can be proved that any two models of the theory are indistinguishable apart from the terminology and notation they employ. A theory which has essentially only one model is called categorical. This notion can be made precise for theories having structures as models in terms of the concept we discuss next.

Let

$$\mathfrak{A} = \langle A, \langle R_i \rangle_{i<k}, \langle F_i \rangle_{i<l}, \langle a_i \rangle_{i<m} \rangle$$

and

$$\mathfrak{B} = \langle B, \langle S_i \rangle_{i<k}, \langle G_i \rangle_{i<l}, \langle b_i \rangle_{i<m} \rangle$$

be structures of the same signature

$$\sigma = \langle k,l,m, \langle m_i \rangle_{i<k}, \langle n_i \rangle_{i<l} \rangle.^*$$

Then $\mathfrak{A}$ is **isomorphic** to $\mathfrak{B}$ iff there exists a one-to-one structure-preserving map of A onto B; that is, there exists a bijection $f: A \to B$ such that

(i) for each $i < k$ and for $x_1, x_2, \ldots, x_{m_i} \in A$,

$$\langle x_1, x_2, \ldots, x_{m_i} \rangle \in R_i \text{ iff } \langle f(x_1), f(x_2), \ldots, f(x_{m_i}) \rangle \in S_i,$$

(ii) for each $i < l$ and for $x_1, x_2, \ldots, x_{n_i} \in A$,

$$f(F_i(x_1, x_2, \ldots, x_{n_i})) = G_i(f(x_1), f(x_2), \ldots, f(x_{n_i})),$$

(iii) for each $i < m$, $f(a_i) = b_i$.

An instance of this definition is that of isomorphism for partially ordered sets given in §1.10. There it is clear from the phrasing that isomorphism of partially ordered sets is a symmetric relation. It follows easily from the general definition above that isomorphism of structures is a symmetric (as well as a reflexive and transitive) relation and so, if structure

---

* The assumption that the structures have the same signature is entirely natural, since we are interested in comparing only models of some one theory.

$\mathfrak{A}$ is isomorphic to structure $\mathfrak{B}$, we may say simply that $\mathfrak{A}$ and $\mathfrak{B}$ are isomorphic.

An axiomatic theory defined in terms of a structure is called **categorical** iff any two models of it are isomorphic. An example of a categorical theory is obtained by adding to the axioms for the group theory the following.

G_0.   The set G has exactly two members.

The resulting theory is consistent by virtue of the model given in Example 2.8.1. The proof that it is categorical is left as an exercise.

Analogous to the acceptance of the existence of a model as a criterion for consistency, the existence of essentially only one model (that is, categoricity) is often accepted as a criterion for negation completeness. To state the pertinent result we make a definition. A statement of a consistent theory $\mathfrak{T}$ will be called a **consequence of** $\mathfrak{T}$ if it is true in every model of $\mathfrak{T}$. Then, *if $\mathfrak{T}$ is a consistent and categorical theory, for each $\mathfrak{T}$-statement S, either S is a consequence of $\mathfrak{T}$ or $\sim$S is a consequence of $\mathfrak{T}$.* This, it will be noted, amounts to negation completeness with provability replaced by a weaker notion. The proof makes use of the following property of models. If $\mathfrak{M}_1$ and $\mathfrak{M}_2$ are isomorphic models of a theory $\mathfrak{T}$, then for every $\mathfrak{T}$-statement S, either S is true in both $\mathfrak{M}_1$ and $\mathfrak{M}_2$ or S is false in both. Assuming this as proved, the main result can be derived as follows. Suppose that the $\mathfrak{T}$-statement S is not a consequence of the consistent theory $\mathfrak{T}$. Then, by the definition of consequence, there exists a model $\mathfrak{M}_1$ of $\mathfrak{T}$ which does not satisfy S. Let $\mathfrak{M}$ be any model of $\mathfrak{T}$. Then, since $\mathfrak{M}$ is isomorphic to $\mathfrak{M}_1$, S is not true in $\mathfrak{M}$, and, hence $\sim$S is true in $\mathfrak{M}$. Since $\mathfrak{M}$ is any model of $\mathfrak{T}$, this means that $\sim$S is a consequence of $\mathfrak{T}$.

A theory which is consistent and noncategorical has essentially different (that is, nonisomorphic) models. This is precisely what should be anticipated for a theory intended to axiomatize the common part of several different theories. The theory of groups is an excellent example. Because it has such a general character it has a wide variety of models, which means that it has a wide range of application.

We conclude this section with several miscellaneous remarks. The first involves assigning a precise meaning to the word "formulation" which we have used frequently. As we described it, an informal theory $\mathfrak{T}$ includes a list T_0 of undefined terms, a list T_1 of defined terms, a list P_0 of axioms, and a set P_1 of all those other statements which can be inferred from P_0 in accordance with some system of logic. The set T_0

serves to generate $T_0 \cup T_1$, the set of all technical terms of $\mathfrak{T}$; the set P_0 serves to generate $P_0 \cup P_1$, the set of all theorems of $\mathfrak{T}$. For the ordered pair $\langle T_0, P_0 \rangle$ we propose the name of a "formulation" for $\mathfrak{T}$. A study of $\mathfrak{T}$ may very well culminate in the discovery of other useful formulations. To obtain one amounts to the determination of: (i) a set T_0' which is a subset of $T_0 \cup T_1$ (which may or may not differ from T_0), and (ii) a subset P_0' of $P_0 \cup P_1$ whose member statements are expressed in terms of the members of T_0' and from which the remaining theorems of the theory can be derived. For a pair of the form $\langle T_0', P_0' \rangle$ to be a formulation of $\mathfrak{T}$, it is clearly sufficient that the members of T_0 can be defined by means of those in T_0' and that the statements of P_0 can be derived from those of P_0'. For many of the well-known axiomatic theories there exists a variety of formulations. This is true, for example, of the theory of Boolean algebras discussed in Chapter 4. A rather trivial example appears in §1.11, and we may rephrase it to suit our present purposes: As a different formulation of the theory of partially ordered sets, we may take that consisting of a set X together with a relation that is irreflexive and transitive on X. Another example is implicit in a remark made in §3.1; rephrased, it amounts to the assertion that Hilbert and Pieri gave different formulations of a theory which axiomatizes intuitive plane geometry. Depending on the criteria adopted, one may show a marked preference for one formulation of a theory over others. Aesthetic considerations may influence one's judgment, and the simplicity of the set of axioms in conjunction with the elegance of the proofs may also play an important role. One may prefer a particular formulation because he feels it has a "naturalness" that others lack. He may favor a formulation which involves the fewest number of primitive notions or axioms.

A notion which is pertinent to a formulation of an informal theory is that of the independence of the set of axioms. A set of axioms is **independent** if the omission of any one of them causes the loss of a theorem; otherwise it is **dependent**. A particular axiom (considered as a member of the set of axioms of some formulation) is **independent** if its omission causes the loss of a theorem; otherwise it is **dependent**. Clearly, an independent axiom cannot be proved from the others of a set of which it is a member, and conversely. Further, the set of axioms of a formulation is independent iff each of its members is independent. Models may be used to establish the independence of axioms. For example, the independence of the axioms O_1, O_2, O_3 for the theory of partially ordered sets (see §3.2) may be shown by constructing a model of each of the three theories having exactly two of O_1, O_2, and O_3 as axioms and in which the interpretation of the missing axiom is false. Otherwise expressed, the independence of O_3, for example, is equivalent to the consistency of the theory having O_1, O_2, and the negation of O_3 as axioms. The independence of a set of axioms is a matter of elegance. A dependent set simply contains one or more redundancies; this has no effect on the theory involved.

The foregoing concepts of independence for both individual axioms and sets of axioms have analogues for primitive terms. A given primitive term (considered as a member of the set of primitive terms in a formulation of a theory) is **independent** if it cannot be formulated by means of the remaining primitive terms, and a set of primitive terms is independent if each of its members is independent. Models are also used to show such independence in the following way. To prove that a particular primitive symbol Q of some formulation of a theory $\mathfrak{T}$ is independent of the remaining primitives, we exhibit two models $\mathfrak{M}_1$ and $\mathfrak{M}_2$ of $\mathfrak{T}$ which have the same domain and in which the interpretation of each primitive term except Q is the same, but which give different interpretations to the symbol Q. The following example offers an indication of the reasoning which justifies this conclusion. In the exercises for §3.2 is a formulation of the theory of simply ordered commutative groups. We will show that the binary relation $\leq$ is an independent primitive. For this we introduce the interpretations $\mathfrak{M}_1$ and $\mathfrak{M}_2$, in both of which we take G as $\mathbb{Z}$, $+$ as ordinary addition, and 0 as zero; in $\mathfrak{M}_1$ we take $\leq$ to be the familiar relation of less than or equal to, whereas in $\mathfrak{M}_2$ we take $\leq$ to be the familiar relation of greater than or equal to. Then, clearly, the interpretations of $\leq$ are different (for example, $2 \leq 3$ is true in $\mathfrak{M}_1$ but false in $\mathfrak{M}_2$). We conclude that $\leq$ cannot be defined in terms of the remaining primitives, for otherwise its interpretation would have to be the same in both models since the other primitives are the same.

In order to motivate the final remark, we recall Theorem 1.11.1, which asserts that every partially ordered set is isomorphic to a collection of sets partially ordered by inclusion. That is, to within isomorphism, all models of the theory of partially ordered sets are furnished by collections of sets. In general, a theorem to the effect that for a given axiomatic theory $\mathfrak{T}$ a distinguished subset of the set of all models has the property that every model is isomorphic to some member of this subset is a **representation theorem** for $\mathfrak{T}$. Analogous to the case of the theory of partially ordered sets, where from the outset collections of sets constitute distinguished models, in the case of an arbitrary theory $\mathfrak{T}$, even though it is noncategorical, one particular type of model may seem more natural. In this event a **representation problem** arises—the question whether there can be proved a representation theorem for $\mathfrak{T}$ which asserts that this type of model yields all models to within isomorphism. When such a problem is answered in the affirmative, new theorems may follow for $\mathfrak{T}$ by imitating proof techniques that have proved useful in those theories which, in effect, supply all models.

Exercises

3.3.1. (a) Establish the consistency of the theory of partially ordered sets by way of a model.

(b) Show that this is a noncategorical theory.

(c) Show that the set of axioms $\{O_1, O_2, O_3\}$ for partially ordered sets is independent.

3.3.2. (a) Show that the theory of groups is noncategorical.

(b) Defining a group as an ordered triple $\langle G, \cdot, e \rangle$ such that G_1, G_2, and G_3 of Example 3.2.1 hold, establish the independence of $\{G_1, G_2, G_3\}$. (Suggestion: Use a multiplication table for displaying the operation which you introduce into any set.)

3.3.3. Consider the axiomatic theory having as its primitive terms two sets A and $\mathfrak{B}$ and having as axioms the following.

(i) Each element of $\mathfrak{B}$ is a two-element subset of A.

(ii) If a, a' is a pair of distinct elements of A, then $\{a, a'\} \in \mathfrak{B}$.

(iii) $A \notin \mathfrak{B}$.

(iv) If B, B' is a pair of distinct elements of $\mathfrak{B}$, then $B \cap B' \subseteq A$.

Show that this theory is consistent. Is it categorical?

3.3.4. Consider the axiomatic theory whose primitive terms are a nonempty A and a binary operation $\langle x, y \rangle \mapsto x - y$ (that is, we write the image of $\langle x, y \rangle$ as $x - y$) in A, which satisfies the identity

$$y = x - [(x - z) - (y - z)].$$

Show that this theory is consistent.

3.3.5. Consider the axiomatic theory whose primitive terms are a nonempty set A, a binary operation $\langle x, y \rangle \mapsto x \times y$ in A, and a unary operation $x \mapsto x'$ in A. The axioms are the following.

(i) $\times$ is an associative operation.

(ii) $(x \times y)' = y' \times x'$.

(iii) If $x \times y = z \times z'$ for some z, then $x = y'$.

(iv) If $x = y'$, then $x \times y = z \times z'$ for all z.

(a) Show that the theory is consistent.

(b) Show that this set of axioms is dependent.

3.3.6. Assume that of two isomorphic models of the theory considered in Exercise 3.3.4, one is a group. Prove that the other is a group.

3.3.7. The set $\{e,a,b,c\}$ together with the operation defined by the following multiplication table is a group. Determine six isomorphisms of this group with itself.

	e	a	b	c
e	e	a	b	c
a	a	e	c	b
b	b	c	e	a
c	c	b	a	e

3.3.8. Devise a definition of isomorphism for systems consisting of a set together with two operations.

3.3.9. Consider an axiomatic theory $\mathfrak{T}$ formulated in terms of two sets, whose members are called *points* and *lines*, respectively, and whose axioms are as follows.

(i) Each line is a nonempty set of points.
(ii) The intersection of two lines is a point.
(iii) Each point is a member of exactly two lines.
(iv) There are exactly four lines.

(a) Show that $\mathfrak{T}$ is a consistent theory.
(b) Show that there are exactly six points in a model of $\mathfrak{T}$.
(c) Show that each line consists of exactly three points.
(d) Find two models of $\mathfrak{T}$.
(e) Is $\mathfrak{T}$ categorical? Give reasons for your answer.

3.3.10. Show that the axiomatic theory defined in Exercise 3.3.4 is a formulation of the theory of commutative groups.

3.3.11. Show that the axiomatic theory defined in Exercise 3.3.5 is a formulation of the theory of groups.

3.3.12. Show that the following is another formulation of the theory of groups. A *group* is an ordered triple $\langle G, \cdot, ' \rangle$ such that G is a set, $\cdot$ is a binary operation in G, $'$ is a unary operation in G, and

(i) G is nonempty;
(ii) $\cdot$ is associative;
(iii) $a'(ab) = b = (ba)a'$ for all a and b.

3.3.13. Show that the following is another formulation of the theory of groups. A group is an ordered triple $\langle G, \cdot, e \rangle$ such that G is a set, $\cdot$ is a binary operation in G, e is a member of G, and

(i) $\cdot$ is an associative operation;
(ii) for each a in G, $ea = a$, and there exists a' in G such that $a'a = e$.

3.3.14. Consider the theory whose primitive terms are a set X and a binary operation $\cdot$ in X, and whose axioms are the following.

(i) X is nonempty.
(ii) $\cdot$ is an associative operation.
(iii) To each element a in X there corresponds an element e of X

such that $ea = ae = a$, and a possesses an inverse a' relative to e in X (that is, $aa' = a'a = e$).

Show that if $\langle S, \cdot \rangle$ is a model of the theory, then there exists a partition of S such that each member set determines a group.

3.3.15. Consider the theory $\mathfrak{T}$ whose primitive terms are the power set of a set S and a mapping f on $\mathcal{P}(S)$ into itself, and whose axioms are as follows:

    (i) For all X in $\mathcal{P}(S)$, $X^f \supseteq X$.

    (ii) For all X in $\mathcal{P}(S)$, $(X^f)^f = X^f$.

    (iii) For all X and Y in $\mathcal{P}(S)$, $X \supseteq Y$ implies $X^f \supseteq Y^f$.

Show that another formulation of $\mathfrak{T}$ results on adopting as the sole axiom:

$$(X \cup Y)^f \supseteq (X^f)^f \cup Y^f \cup Y, \text{ for all } X \text{ and } Y \text{ in } \mathcal{P}(S).$$

§3.4. Formal Axiomatic Theories

Let us review the steps we have discussed so far in the formalization of a mathematical theory M. One begins by arranging the known theorems of M in a logical order—if the proof of theorem T requires the statement of theorem S, then theorem T is placed after theorem S. In such an arrangement, there must be initial statements which are not proved on the basis of previous theorems. These are declared to be axioms of the theory. Also needed is an arrangement of the technical terms of the theory to arrive at a set of technical terms such that all others can be defined in terms of these. The members of this set are labeled as the undefined terms and, if need be, additional axioms are supplied to express all properties of the agreed-upon undefined terms which matter for the deductions of the theorems. Then one can carry out the deductions, treating the undefined terms as words having no meaning. If one stops the axiomatization at this point, the result is an informal axiomatic theory as we have discussed this notion. In such a theory, theorems and their proofs are usually stated as sentences in some natural word languages (possibly with some symbolization, for convenience), logical terms in these sentences have their everyday meaning, and the logical principles used in deductions are those acceptable in everyday mathematics.

The formalization of M can be carried further. First, the meaning of all logical terms can be suppressed by stating all conditions which govern their use. Further, the logical principles can be given in part by further axioms and, in part, by rules which permit the inference of a sentence from one or more others. The result is a complete abstraction from the content of M, leaving only its form. We say that (by this process) M has been formalized and call its abstraction a formal theory. In order to give a more precise description of a formal theory, we introduce an auxiliary notion. An **algorithm** or **effective procedure** for a class of questions is a finite set of instructions that provides a mechanical procedure by which the answer to any one of the questions can be found in a finite number of steps. An effective procedure is like a recipe, in that it tells what to do at each step and no intelligence is required to follow it. In more detail, assume that an effective procedure is available for a class of questions. Then, if we select any question of the class, the procedure will tell us how to perform successive steps, after a finite number of which we will have the answer. After any step, if the answer has not been obtained, the instructions, together with the existing situation in the computation, will tell us what to do next; that is, to perform the next step requires no intelligence or insight on our part. Finally, the instructions will enable us to recognize when the last step has been completed and what the answer is. In principle, it is possible to construct a machine for executing the instructions of an algorithm.

Some classes of questions call for a "yes" or "no" answer; examples are the class of questions

1.   Is the statement form A a tautology?
and the class
2.   Is the integer a a factor of the integer b?
Other classes of questions require the determination of some object; examples are
3.   What is the greatest common divisor of the integers a and b?
and
4.   What is the sum of the positive integers a and b?

From a theoretical point of view, there is no interest in an effective procedure for a finite class of questions, each of which has a "yes" or "no" answer, since a list of answers to the individual questions could be prepared and serve as an algorithm. On the other hand, there is considerable interest in the existence of an effective procedure for a denumerable class of questions; an algorithm for such a class is a finite set of instructions that will provide the answer to any one of the infinitely many questions in that class. For each of the denumerable classes described above there is an algorithm. The method of truth tables provides an algorithm for (1) and long division suffices for (2). The classical Euclidean algorithm is an effective procedure for (3). An algorithm for (4) is learned in grade school.

It should be recognized that we have not given a *definition* of an effective procedure, but only an intuitive description. This suffices so long as we limit ourselves to positive assertions that an effective procedure exists for a certain class of questions and document the statement with a description of the procedure and an argument that it is effective. On the other hand, to prove a negative result, that for a certain class of questions no effective procedure exists, requires a precise definition of the concept "effective procedure." In the 1930's general agreement on such a definition was reached. It is touched on in §3.7, and is presented in some detail in the Appendix to this chapter.

Of particular importance to us is the existence of effective procedures for deciding whether a given object is a member of a specified set. A set is called (**effectively**) **decidable** if there is an algorithm for deciding whether any object is a member of that set.

We now turn to the definition which has been promised. A **formal theory** $\mathfrak{T}$ consists of four related sets. The first of these is a finite or denumerable set of symbols called the **alphabet** of $\mathfrak{T}$. The second, whose members are called the **well-formed formulas** (abbreviated "wfs") of $\mathfrak{T}$, is a subset of the set of all finite sequences of elements of the alphabet. (A finite sequence of elements of the alphabet of $\mathfrak{T}$ is called an **expression** of $\mathfrak{T}$). The third set, whose elements are called the **axioms** of $\mathfrak{T}$, is a subset of the set of wfs. The remaining set is a finite set $\{R_1, \ldots, R_m\}$ where R_i is an n_i-ary relation in the set of wfs. Each R_i is called a **rule of inference** of $\mathfrak{T}$.

The definition of $\mathfrak{T}$ is completed with the addition of the following requirements.

(i) The alphabet of $\mathfrak{T}$ is a decidable set.

(ii) There is an effective procedure to decide whether a given expression is a wf.

(iii) If the n-ary relation R in the set of wfs of $\mathfrak{T}$ is a rule of inference, then there exists an effective procedure to decide whether or not a given n-tuple of wfs is in R.

If the set of axioms of a formal theory is decidable, then it is called an **axiomatic theory.**

Associated with a given formal theory is a deductive structure, which we describe next. Let $\mathfrak{T}$ be such a theory and X be a set of wfs of $\mathfrak{T}$. We shall say that a wf A is **directly deducible** from X if there is an n-ary rule of inference R and a finite sequence $B_1, \ldots, B_{n-1}$ of wfs in X such that $\langle B_1, \ldots, B_{n-1}, A \rangle \in R$. A **proof** in $\mathfrak{T}$ is a sequence $A_1, \ldots, A_r$ of wfs of $\mathfrak{T}$ such that, for each i, either A_i is an axiom of $\mathfrak{T}$ or A_i is directly deducible from some preceding wfs (by virtue of one of the rules of inference). A **theorem** of $\mathfrak{T}$ is a wf A of $\mathfrak{T}$ such that there is a proof which has A as its last member. Such a proof is called a **proof of** A. If $\mathfrak{T}$ is a formal *axiomatic* theory, the notion of proof is effective—given any finite sequence of wfs, we can decide whether or not it is a proof. It does *not* follow, even if $\mathfrak{T}$ is axiomatic, that the set of theorems is decidable. A theory for which the theorems constitute a decidable set is called a **decidable theory.** A decidable theory is, roughly speaking, one for which a machine can be devised to test wfs for theoremhood. In contrast, in undecidable theories (which include many interesting theories) ingenuity is required to determine whether some wfs are theorems.

A wf A is said to be **deducible** in $\mathfrak{T}$ from a set X of wfs iff there exists a sequence $A_1, \ldots, A_n$ of wfs such that $A_n = A$ and, for each i, either A_i is an axiom, or A_i is in X, or A_i is directly deducible from some of the preceding wfs in the sequence. Such a sequence is called a **deduction** of A from X.* The members of X are called the **hypotheses** or **premises** of the proof. We shall abbreviate "A is deducible in $\mathfrak{T}$ from X" by

$$X \vdash_{\mathfrak{T}} A.$$

If there is no need to mention the theory, the subscript $\mathfrak{T}$ will be omitted. If X is a finite set $\{B_1, \ldots, B_n\}$, we shall write $B_1, \ldots, B_n \vdash A$ instead of $\{B_1, \ldots, B_n\} \vdash A$. When X is empty, we write simply $\vdash A$ instead of $\varnothing \vdash A$. Clearly, $\vdash A$ means that A is a theorem. The reader may verify the following simple properties of the notion of deducibility.

(i) $X \vdash A$ whenever $A \in X$ or A is an axiom.
(ii) If $X \vdash A$ and Y is any set of wfs, $X \cup Y \vdash A$.
(iii) If $X \vdash A$, there is a finite subset Y of X such that $Y \vdash A$.
(iv) If $Y \vdash A$ and, for each B in Y, $X \vdash B$, then $X \vdash A$.

§3.5. The Statement Calculus as a Formal Axiomatic Theory

In view of the role of the statement calculus in a theory of inference (§2.4), the goal of an axiomatization is a formal axiomatic theory in which the theorems are precisely the tautologies. This was first achieved in Frege [1879]. Since then many formulations have appeared. That

---

* Note that if X is empty, then a deduction of A from X is a proof of A.

which we shall present is the simplification of Frege's formulation devised by Łukasiewicz. The alphabet of the theory L to be defined consists of the symbols

$$\sim, \rightarrow, (\ ),$$
$$\mathcal{A}, \mathcal{B}, \mathcal{C}, \mathcal{A}_1, \mathcal{B}_1, \mathcal{C}_1, \ldots .$$

The symbols $\sim$ and $\rightarrow$ are called **primitive connectives** and those in the second line are called **statement letters.** The three dots, which are not symbols, indicate that the list continues without end. The wfs are defined as follows.

(a) A statement letter is a wf.
(b) If A and B are wfs, so are $(\sim A)$ and $(A \rightarrow B)$.
(c) An expression is a wf only if it can be shown to be so on the basis of (a) and (b).

If A, B, and C are any wfs of L, then each of the following wfs is an axiom of L.

(L1) $(A \rightarrow (B \rightarrow A))$
(L2) $((C \rightarrow (A \rightarrow B)) \rightarrow ((C \rightarrow A) \rightarrow (C \rightarrow B)))$
(L3) $(((\sim B) \rightarrow (\sim A)) \rightarrow (A \rightarrow B))$

The only rule of inference is modus ponens: B is directly deducible from A and $A \rightarrow B$.

Clearly, the alphabet of L is decidable. An algorithm for determining whether an expression is a wf can be formulated: its ingredients are a rule for pairing left and right parentheses in an expression and a rule for methodically examining the expression within a matched pair of parentheses to determine whether it constitutes a wf. Further, there is an effective procedure to decide whether a given wf is directly deducible from a given pair of wfs.

Since the axioms are written in terms of arbitrary wfs, each stands for infinitely many axioms. Thus, each of (L1) to (L3) is referred to as an **axiom schema.** Even though there are infinitely many axioms, the set is decidable, since each axiom must have one of three forms. Hence, L is axiomatic.

We introduce other connectives by definition as in §2.7:

$(A \wedge B)$ stands for $(\sim(A \rightarrow (\sim B)))$;
$(A \vee B)$ stands for $((\sim A) \rightarrow B)$;
$(A \leftrightarrow B)$ stands for $((A \rightarrow B) \wedge (B \rightarrow A)).$*

---

* Hereafter we shall omit the outermost pair of parentheses of wfs and invoke the conventions introduced earlier for the elimination of parentheses.

An illustration of a formal proof is given next. It is a proof of the formula $\alpha \to \alpha$. It follows that $\vdash \alpha \to \alpha$.

(1) $(\alpha \to ((\mathcal{B} \to \alpha) \to \alpha)) \to ((\alpha \to (\mathcal{B} \to \alpha)) \to (\alpha \to \alpha))$

                                                                    Axiom schema (L2)
(2) $\alpha \to ((\mathcal{B} \to \alpha) \to \alpha)$ Axiom schema (L1)
(3) $(\alpha \to (\mathcal{B} \to \alpha)) \to (\alpha \to \alpha)$ 1, 2 modus ponens
(4) $\alpha \to (\mathcal{B} \to \alpha)$ Axiom schema (L1)
(5) $\alpha \to \alpha$ 3, 4 modus ponens

When a proof is given, an analysis (that is, an explanation to justify each occurrence of a wf in it) is usually given in parallel, as above. This is not required, however, because there is an effective procedure for supplying an analysis.

We observe that we can just as easily prove $\vdash \mathcal{B} \to \mathcal{B}$ or $\vdash (\mathcal{C} \wedge \alpha) \to (\mathcal{C} \wedge \alpha)$ by repeating the above sequence of formulas with $\mathcal{B}$ or $\mathcal{C} \wedge \alpha$ in place of α. Indeed, if in the above formal proof we substitute any wf A for the statement variable "α," we get a formal proof of the wf $A \to A$. But if, instead, we substitute the variable "A" for "α" (and, "B" for "$\mathcal{B}$") we get a **proof schema** of the **theorem schema** "$A \to A$." A theorem schema, like an axiom schema, has the merit that a theorem results when the same wf is chosen for all occurrences of any letter that appears in it.

That the proof of such a simple wf as $A \to A$ is so long may cause alarm. Generally the development of a formal theory is tedious at the outset since there is little to work with. Often relief can be found in the form of theorems, called **derived rules of inference,** which assert the existence of proofs or deductions under various conditions. An example of such a rule for the statement calculus is

$$\text{If } \vdash A \text{ and } \vdash A \to B, \text{ then } \vdash B.$$

An application of this derived rule, whose proof is left as an exercise, is commonly called "modus ponens" because of its similarity to the primitive rule of inference bearing that name. Another derived rule which plays a more significant role in obtaining theorems of L is the following.

THEOREM 3.5.1. (The Deduction Theorem). If X is a set of wfs of L, and A and B are wfs of L, then

$$X, A \vdash B \text{ implies } X \vdash A \to B$$

(using $X, A \vdash B$ to stand for $X \cup \{A\} \vdash B$).

Proof. Let $B_1, \ldots, B_n(=B)$ be a deduction of B from $X \cup \{A\}$. We shall prove by induction on m that $X \vdash A \to B_m$ for $1 \leq m \leq n$. Now B_1 must be in X, or an axiom of L, or A. By axiom (L1), $B_1 \to (A \to B_1)$ is an axiom. Hence, in the first two cases, by modus ponens, $X \vdash A \to B_1$. In the third case, when B_1 is A, we have $\vdash A \to B_1$ (this was proved above), and so $X \vdash A \to B_1$. This disposes of the case $m = 1$. Next assume that $X \vdash A \to B_r$ for $r \leq m$. Either B_{m+1} is an axiom, or is in X, or is A, or follows by modus ponens from some B_j and B_k where $j, k \leq m$ and B_k has the form $B_j \to B_{m+1}$. In the first three cases, $X \vdash A \to B_{m+1}$ as in the case $m = 1$ above. In the last case we have, by the induction hypotheses, $X \vdash A \to B_j$ and $X \vdash A \to (B_j \to B_{m+1})$. Further, by axiom (L2),

$$\vdash (A \to (B_j \to B_{m+1})) \to ((A \to B_j) \to (A \to B_{m+1})).$$

Two applications of modus ponens yield $X \vdash A \to B_{m+1}$ and the proof is complete. The case $m = n$ is the desired result.

Instead of illustrating the usefulness of the deduction theorem in deriving theorems of L, we employ it in another connection. With repeated applications of the theorem, it follows that if $A_1, A_2, \ldots, A_m \vdash B$, then $\vdash A_1 \to (A_2 \to (\cdots(A_m \to B)\cdots))$. The converse of this assertion also holds; the proof is left as an exercise. These two results may be summarized as follows.

THEOREM 3.5.2. For wfs $A_1, A_2, \ldots, A_m$ and B of L,

$$A_1, A_2, \ldots, A_m \vdash B \text{ iff } \vdash A_1 \to (A_2 \to (\cdots(A_m \to B)\cdots)).$$

This theorem, which reduces the syntactical notion of deducibility to provability, parallels the result in the corollary to Theorem 2.4.1, which reduces the semantic notion of consequence to validity. It follows that if a wf is a theorem iff it is a tautology, the equivalence of the informal and the formal statement calculus, both by themselves and when applied under a set of assumption formulas, will be demonstrated. The theorem which asserts that every tautology is a theorem is an example of a completeness theorem in the positive sense as discussed in §3.3. It was first derived in 1921 by the American logician Emil Post. The proof, although elementary, is tedious and on this account will not be given.*

THEOREM 3.5.3. (The Completeness Theorem). If A is a tautology, then A is a theorem; in brief, if $\models A$ then $\vdash A$.

* A proof may be found in Stoll (1963), Chapter 9.

In contrast, the converse statement is established easily, as we show next.

THEOREM 3.5.4. If A is a theorem of L, then A is a tautology; that is, if $\vdash A$ then $\models A$.

Proof. We observe first that each instance of an axiom schema is a tautology; that is, the theorem is true if A is an axiom. Further, by Theorem 2.3.3, if $\models A$ and $\models A \to B$, then $\models B$. Since every theorem is either an axiom or comes from the axioms by one or more uses of modus ponens, every theorem is a tautology.

Taken together, Theorems 3.5.3 and 3.5.4 establish that the notions of provability and validity are coextensive for L. Since there is an algorithm for deciding whether a wf of L is a tautology, L is a decidable theory.

Examples

3.5.1. From Theorems 3.5.2 to 3.5.4, derived rules of inference can be justified by appropriate tautologies. Below are listed five such rules, with the tautology that justifies each one placed opposite it.

$$A \vdash A \lor B \qquad\qquad \models A \to (A \lor B)$$
$$A \lor B, \sim A \vdash B \qquad\qquad \models A \lor B \to (\sim A \to B)$$
$$A, B \vdash A \qquad\qquad \models A \to (B \to A)$$
$$\sim B \to \sim A \vdash A \to B \qquad\qquad \models (\sim B \to \sim A) \to (A \to B)$$
$$\sim B \to \sim A, A \vdash B \qquad\qquad \models (\sim B \to \sim A) \to (A \to B)$$

3.5.2. As an illustration of the incorporation of a theory of inference in a formal theory, which is a feature of a formal theory, we outline the imbedding of the statement calculus into a formal theory. This may be accomplished by

(i) adding to the alphabet any of the symbols $\sim$, $\to$, (, and) which are not already present;

(ii) including among the formation rules for wfs the following:

If A and B are wfs, then so is $(A \to B)$,
If A is a wf, then so is $(\sim A)$;

(iii) adding to the axioms of the theory the three axiom schemas we have chosen for the theory L (where now "wf" is taken in the extended sense of "formula of the theory");

(iv) adding modus ponens to the rules of inference.

Formulas of the theory may then be regarded as wfs of a statement calculus in which the role of the statement letters is played by those wfs which are not of the form $(A \to B)$ or $(\sim A)$, that is, wfs which cannot be decomposed into further wfs using $\to$ and $\sim$ in the way shown.

As a result of the imbedding, every tautology will be a theorem of the theory. More important, the statement calculus is available as a theory of inference. This theory is adequate to provide the logical skeleton of various kinds of proofs that are encountered frequently. A few examples follow.

(a) To establish that a wf B of a theory, into which the statement calculus is incorporated, is a theorem, it is sufficient to prove that $\sim B \to \sim A$ and A are theorems. This follows from the last derived rule of inference in Example 3.5.1. Similarly, the rule $\sim B \to \sim A \vdash A \to B$ justifies a proof by contraposition.

(b) Let us use "C" to denote a contradiction. In formal terms, the proof of a wf A by contradiction may be stated as

$$\text{If } \sim A \vdash C, \text{ then } \vdash A.$$

This rule stems from the tautology $(\sim A \to C) \to A$. In practice, such a proof may take the following form. One shows that $\sim A \vdash B$ and $\vdash \sim B$ and infers that $\sim A \vdash B \wedge \sim B$, and then $\vdash A$.

(c) To establish that a conditional $A \to B$ is a theorem with a proof by contradiction, the following rule is often used:

$$\text{If } A, \sim B \vdash C \text{ (a contradiction), then } \vdash A \to B.$$

The reader may justify this.

(d) A "proof by cases" is not uncommon in mathematics. Such a proof of a wf B begins with the enumeration of a finite set $A_1, A_2, \ldots, A_m$ of wfs which are exhaustive in the sense that $\vdash A_1 \vee A_2 \vee \cdots \vee A_m$. Then proofs of $A_1 \to B, A_2 \to B, \ldots, A_m \to B$ are provided and it is concluded that B is a theorem. The rule at hand is

$$\text{If } \vdash A_1 \vee A_2 \vee \cdots \vee A_m, \vdash A_1 \to B, \ldots, \vdash A_{m-1} \to B,$$
$$\text{and } \vdash A_m \to B, \text{ then } \vdash B.$$

§3.6. First-Order Theories

In §2.8 appeared rigorous definitions of the concepts of interpretation, truth, and logical validity as they pertain to first-order languages. We have stated that, according to a generally agreed-upon definition of "effective," it can be proved that there is no effective way to test for logical validity. That is, there is no algorithm for the class of questions "Is the wf A (of a specified language) logically valid?" This fact provides the motivation for applying the axiomatic method to the study of wfs involving quantifiers. It amounts to the formulation of a first-order language as a formal theory by specifying axioms and rules of inference. When so formulated, a first-order language will be called a **first-order**

theory. The rules of inference will be the same for all theories. Further, there will be a set of axioms common to all theories. Individual first-order theories may have other axioms; those we specify as common to all are called **logical axioms.** They constitute a decidable set and the rules of inference are decidable. The logical axioms together with the rules of inference are called the axioms and rules of the predicate calculus. A first-order theory whose only axioms are the logical axioms is called a first-order **predicate calculus.** As we shall see, the logical axioms are so designed that the logical consequences (in the semantic sense of §2.9) of the set of all axioms of a first-order theory $\mathfrak{T}$ are precisely the theorems of $\mathfrak{T}$. In particular, if $\mathfrak{T}$ is a first-order predicate calculus, it turns out that the theorems of $\mathfrak{T}$ are precisely those wfs of $\mathfrak{T}$ which are logically valid. Thus we can replace the notion of validity by that of provability in predicate calculi.

There is another virtue of the plan proposed above. Although the definition of validity depends on set-theoretical notions, the notion of provability does not. Set theory, because of the paradoxes, is considered an insecure foundation on which to build mathematical logic. Logicians consider the syntactical approach, consisting of a study of formal theories using only elementary number theory and the principle of mathematical induction, as a safer path.

At this point the reader should review §2.7 to refresh his memory of the alphabet, the definitions given there for term, wf, and the connectives $\wedge$, $\vee$, $\leftrightarrow$, and the agreements about the omission of parentheses in wfs, for what we now call a first-order theory. If $\mathfrak{T}$ is any first-order theory, and if A, B, and C are any wfs of $\mathfrak{T}$, then the **logical axioms** of $\mathfrak{T}$ are the following.

1. $A \rightarrow (B \rightarrow A)$
2. $(C \rightarrow (A \rightarrow B)) \rightarrow ((C \rightarrow A) \rightarrow (C \rightarrow B))$
3. $(\sim B \rightarrow \sim A) \rightarrow (A \rightarrow B)$
4. $(x_i)(A \rightarrow B) \rightarrow (A \rightarrow (x_i)B)$, if A is a wf of $\mathfrak{T}$ containing no free occurrence of x_i.
5. $(x_i)A(x_i) \rightarrow A(t)$, if $A(x_i)$ is a wf of $\mathfrak{T}$ that admits the term t for x_i.

A first-order theory may have other axioms; if so, they are called **proper axioms** (meaning proper to the particular theory in question).*
Above we agreed to call a first-order theory having only the logical axioms a predicate calculus. By *the* predicate calculus is meant the

---

* We note that the axioms common to all first-order theories are axiom schemes. In contrast, proper axioms (which are also called **postulates**) are usually not schemes.

theory of deduction based on the logical axioms (1)–(5) above and the rules of inference given below. A **theorem of the predicate calculus** is any theorem of any predicate calculus. The **pure predicate calculus** is the first-order predicate calculus with no function letters, no constant letters, and all the predicate letters.

The rules of inference of any first-order theory are

(i) Modus ponens: B is directly deducible from A and $A \rightarrow B$;

(ii) Generalization: $(x_i)A$ is directly deducible from A.

A **model** of a first-order theory $\mathfrak{T}$ is an interpretation in which all the axioms are true.* According to parts (d) and (f) of Theorem 2.8.2, if the rule of modus ponens or that of generalization is applied to wfs true for an interpretation, then the result of the application is also true. Hence, every theorem of $\mathfrak{T}$ is true in any model of $\mathfrak{T}$.

Examples

3.6.1. The formulation of group theory as a first-order theory follows immediately from the presentation in Example 2.7.3. It is the language defined there augmented with the logical axioms (1)–(5) above and the axioms given in the example (these are the proper axioms) and having modus ponens and generalization as rules of inference. Often such a formulation is referred to as the **elementary theory of groups.** The adjective "elementary" signifies that the first-order predicate calculus is the theory of deduction that is employed.

Another formulation (see Exercise 3.2.7) of group theory as a first-order theory is as follows. The theory has one predicate letter P_1^2, one function letter f_1^2, and no individual constant. To conform with ordinary notation, we shall write $s = t$ for $P_1^2(s,t)$ and $s \cdot t$ for $f_1^2(s,t)$. The proper axioms are

$(x_1)(x_2)(x_3)(x_1 \cdot (x_2 \cdot x_3) = (x_1 \cdot x_2) \cdot x_3)$,
$(x_1)(x_2)(\exists x_3)(x_1 = x_2 \cdot x_3)$,
$(x_1)(x_3)(\exists x_2)(x_1 = x_2 \cdot x_3)$,
$(x_1)(x_1 = x_1)$,
$(x_1)(x_2)(x_1 = x_2 \rightarrow x_2 = x_1)$,
$(x_1)(x_2)(x_3)(x_1 = x_2 \wedge x_2 = x_3 \rightarrow x_1 = x_3)$,
$(x_1)(x_2)(x_3)(x_2 = x_3 \rightarrow (x_1 \cdot x_2 = x_1 \cdot x_3 \wedge x_2 \cdot x_1 = x_3 \cdot x_1))$.

The theory of groups is axiomatic. In general, any theory with a finite number of proper axioms is axiomatic, since it is clear that one can effectively decide whether a given wf is a logical axiom.

3.6.2. From Example 2.7.4, a formulation of the theory of partially ordered sets as a first-order theory can be inferred immediately.

3.6.3. A first-order theory, which will be symbolized by N, for the arithmetic

---

* Since it will be shown (see the proof of Theorem 3.6.2) that the logical axioms are logically valid wfs, they are true in *every* interpretation. Hence, the question of whether an interpretation of $\mathfrak{T}$ is a model rests with the proper axioms.

of natural numbers that is based upon Peano's axioms has one predicate letter P_1^2 (we write $s = t$ for $P_1^2(s,t)$), one individual constant a_1 (we write 0 for a_1), and three function letters f_1^1, f_1^2, f_2^2. We shall write s' for $f_1^1(s)$, $s + t$ for f_1^2, and $s \cdot t$ for $f_2^2(s,t)$. The proper axioms of the theory are

$$(x_1 = x_2) \wedge (x_2 = x_3) \to x_1 = x_3,$$
$$x_1 = x_2 \to x_1' = x_2',$$
$$x_1' \neq 0,$$
$$x_1' = x_2' \to x_1 = x_2,$$
$$x_1 + 0 = x_1,$$
$$x_1 + x_2' = (x_1 + x_2)',$$
$$x_1 \cdot 0 = 0,$$
$$x_1 \cdot x_2' = x_1 \cdot x_2 + x_1,$$

For any wf $A(x)$, $A(0) \wedge (x)(A(x) \to A(x')) \to (x)A(x)$.

Those additional properties of equality that one might expect to be included among the axioms can be derived as theorems in N. The last axiom, which is actually an axiom schema, is a restricted form of the induction axiom (see §1.12). The theory N is often referred to as **first-order arithmetic.**

The corollary of Theorem 2.4.1 reduces the notion of consequence to that of validity for the statement calculus. Remarks made in the beginning of §2.9 imply that this semantic result holds also for first-order theories. Theorem 3.5.2 expresses a parallel result for the statement calculus—the notion of deducibility can be reduced to provability. This result extends to first-order theories, but we shall not prove it.* There comes to mind next the question of the relationship between validity and provability. A far-reaching and deep result in this connection is stated next without proof.†

THEOREM 3.6.1. A logically valid wf of a first-order theory $\mathfrak{T}$ is a theorem of $\mathfrak{T}$.

From this follows the difficult part of a theorem first proved by K. Gödel in 1930.

THEOREM 3.6.2. (Gödel's completeness theorem for a predicate calculus). In a first-order predicate calculus, the theorems are exactly the logically valid wfs.

Proof. If A is a wf of a predicate calculus and $\models A$, then $\vdash A$ according to Theorem 3.6.1. Conversely, suppose that $\vdash A$. According

---

* To state the appropriate theorem correctly requires a detailed formulation of rules governing how any variable having a free occurrence in a premise is to be treated in deductions. A discussion of this may be found in Kleene (1967), Chapter 2.

† A proof is given in Mendelson (1964), Chapter 2.

to Example 2.8.5, axioms (1)–(3) are logically valid. By Exercise 2.8.4 and Theorem 2.8.3(I), axioms (4) and (5) are logically valid. By parts (d) and (f) of Theorem 2.8.2, the rules of inference preserve logical validity. It follows that if $\vdash A$ then $\vDash A$.

Using this result and an extension of the earlier deduction theorem to predicate calculi, the following result is easily proved. Let A be a wf of a first-order theory $\mathfrak{T}$. Then $\vdash_\mathfrak{T} A$ iff there exists a wf C which is the closure of the conjunction of instances of some axioms of $\mathfrak{T}$ such that $C \vDash A$. In other words, the logical consequences of the axioms of $\mathfrak{T}$ coincide with the theorems of $\mathfrak{T}$. It can also be shown that a wf A is true in every model of $\mathfrak{T}$ iff $\vdash_\mathfrak{T} A$. From this result one can derive the following: If X is a set of wfs of $\mathfrak{T}$ and $X \vDash A$, then $X \vdash_\mathfrak{T} A$. The foregoing result together with Theorems 3.6.1 and 3.6.2 show that the syntactical approach to first-order logic by way of first-order theories is equivalent to the semantic approach through the notions of interpretations, truth, models, and logical validity.

§3.7. Metamathematics

The principal reason for formulating intuitive theories as formal axiomatic theories and, in particular, as first-order theories, is that such fundamental notions as consistency and completeness can be discussed in a precise and definitive way. This is possible because the notion of proof is made explicit. Before we turn to theorems related to such matters, we should have some understanding of *how* such matters are studied and *why* such methods are used. In this section we shall describe the admissible methods for the study of formal theories as advocated by the school of formalists (founded by Hilbert) and then prove some theorems in accordance with these methods.

A formal theory is a completely symbolic language built according to certain rules from the alphabet of specified primitive symbols. When a formal theory becomes the object of study it is called an **object language.** To discuss it, which includes defining its syntax, specifying its axioms and rules of inference, and analyzing its properties, another language— the **metalanguage** or syntax language—is employed. Our choice of a metalanguage is the English language. In general terms the contrast between a metalanguage and the object language which is discussed in terms of this metalanguage is parallel to the contrast between the English language and the French language for one whose native tongue is

English and who studies French. At the outset, vocabulary, rules of syntax, and so on, are communicated in English (the metalanguage). Later, one begins to write in French. That is, one forms sentences within the object language. To give a concrete example, consider the elementary theory of groups as formulated in the preceding section. The statement "The elementary theory of groups is an undecidable theory" is about group theory and written in the English language—that is, in the metalanguage. In contrast,

$$"(a)(b)(c)(a \cdot b = a \cdot c \rightarrow b = c)"$$

is a statement of group theory—that is, of the object language.

A theorem *about* a formal theory is called a **metatheorem** and is to be distinguished from a theorem *of* the theory. It is easy to make this distinction, since a theorem of the theory is written in the symbolism of the theory, whereas a metatheorem is written in English. In the preceding paragraph the statement in English regarding group theory is a metatheorem, and that written in terms of $\cdot$, $=$, and so on, is a theorem of group theory. Since the proof of a metatheorem requires a system of logic, a description of the system of logic should be available for the prospective user of the metatheorem. One possibility is to formalize the metalanguage as we have formalized the predicate calculus. But this entails the use of a metametalanguage, and the beginning of an unending regress is established. The alternative, which was proposed by Hilbert, may be summarized roughly: In the metalanguage employ an informal system of logic whose principles are universally accepted.

More generally, Hilbert took the position that a metatheory (that is, the study of a formal theory in the metalanguage selected) should have the following form. First of all, it should belong to intuitive and informal mathematics; thus, it is to be expressible in ordinary language with mathematical symbols. Further, its theorems (that is, the metatheorems of the formal theory) must be understood and the deductions must carry conviction. To help ensure the latter, all controversial principles of reasoning such as the axiom of choice must not be used. More generally, the methods used in the metatheory should be restricted to those called *finitary* by the formalists. This excludes consideration of infinite sets as "completed entities" and requires that an existence proof provide an effective procedure for constructing the object which is asserted to exist. Mathematical induction is admissible as a finitary method of proof, since a proof by induction of the statement "For all n, $P(n)$" shows that any given natural number n has the property expressed by P by reasoning which uses only the numbers from 0 up to n; that is, induction does not require one to introduce the classical completed infinity of the natural numbers. Finally it is assumed that if, for example, the English language is taken as the metalanguage, then only a minimal fragment will be used. (The danger in permitting all of the English language to be used is that one can derive within it the classical paradoxes, for example, Russell's paradox.) By **metamathematics** or **proof theory** is meant the study of formal theories using methods which fit into the foregoing framework. In brief, metamathematics is the study of formal theories by methods which should be convincing to everyone qualified to engage in such activities.

Before discussing some metamathematical notions and proving some metatheorems, we outline the reasons which led Hilbert to formulate metamathematics as he did. The introduction of general set theory with its abstractness and its treatment of notions (such as the completed infinite) which are inaccessible to experience, yet with its fruitful applications to concrete problems of classical mathematics, provided the stimulus for investigations of the foundations of mathematics in the sense that this subject matter is now known. The discovery of contradictions within set theory served to strengthen and accelerate these investigations. The initial reaction to the antinomies of intuitive set theory was a reconstruction of set theory as an axiomatic theory, placing restrictions (as few as appear to be required) around the notion of a set, for the purpose of eliminating the known antinomies.

Some felt that even if this venture should prove to be successful, it would not provide a complete solution to the problem because, they argued, the paradoxes raised questions about the nature of mathematical proofs and criteria for distinguishing between correct and incorrect proofs for which satisfactory answers had not been provided. Russell, for example, judged the cause of the paradoxes to be that each involves an impredicative procedure. This led Russell to formulate a system of logic (his *ramified theory of type*, 1908) in which impredicative procedures are excluded and, with Whitehead, to attempt to develop mathematics as a branch of logic (*Principia Mathematica*). Both the logistic school and the advocates of the axiomatic approach to set theory, initiated by E. Zermelo, were in need of proofs of the consistency of their theories. It was recognized that the classical method of providing a proof—the exhibition of a model within the framework of a theory whose consistency was not in doubt—could not be applied. Further, finite models were clearly inadequate, and no conceptual framework within which an infinite model might be constructed could be regarded as "safe" in view of the antinomies. It was Hilbert who contributed the idea of making a direct attack upon the problem of consistency by proving *as a theorem about* each such theory that contradictions could not arise. Hilbert recognized that in order to carry out such a program, theories would have to be formalized so that the definition of *proof* would be entirely explicit. To this end he brought the notion of a formal axiomatic theory to its present state of perfection. To prove theorems about such theories—in particular, to attack the problem of consistency —Hilbert devised metamathematics. By restricting the methods of proof to be finitary in character, he hoped to establish the consistency of theories such as *N* with the same degree of impeachability as is provided by proofs of consistency via finite models when the latter technique is possible (as in group theory, for instance).

For our first example of a metamathematical notion we choose *consistency*. The definition in §3.3 (a theory is consistent iff for no formula *A* both *A* and $\sim A$ are provable) is applicable to any formal theory having the ·symbol $\sim$ for negation. It is metamathematical since it refers only to the formal symbol $\sim$ and the definitions of formula and provability. A metatheorem concerning a class of theories to which the definition is applicable is proved next.

THEOREM 3.7.1.   Let $\mathfrak{T}$ be a formal theory which includes the statement calculus. Then $\mathfrak{T}$ is consistent iff not every formula of $\mathfrak{T}$ is a theorem.

Proof.   Suppose that $\mathfrak{T}$ is inconsistent and that *A* is a wf such that both $\vdash A$ and $\vdash \sim A$. Now $A \rightarrow (\sim A \rightarrow B)$ is a theorem for any *B* since it is a tautology. Hence *B* (that is, any wf) is a theorem by two applications of modus ponens. For the converse, assume that every wf of $\mathfrak{T}$ is a theorem. Then if *A* is any wf, both *A* and $\sim A$ are theorems. Thus, $\mathfrak{T}$ is inconsistent.

Henceforth it will be assumed that all formal theories include the statement calculus so that Theorem 3.7.1 will always hold. Our next result is a metatheorem about the statement calculus.

THEOREM 3.7.2.   The statement calculus is a consistent theory.

Proof.   Let *A* be a theorem. Then, in turn, *A* is a tautology, $\sim A$ is not a tautology, and $\sim A$ is not a theorem.

The foregoing is a metamathematical proof. To substantiate this assertion, we note first that the computation process for filling out a truth table for a wf is metamathematical. Hence the property of being a tautology is a metamathematical property of wfs of the statement calculus. It follows that the proof of Theorem 3.5.4 is metamathematical. Since the proof in question relies solely on Theorem 3.5.4, it also is metamathematical.

A similar argument using Theorem 3.6.2 yields a proof of the consistency of any predicate calculus $\mathfrak{T}$, as soon as a wf which is not valid is exhibited. This proof is not metamathematical because the notion of validity is not. However, a metamathematical proof can be given*; the idea is to map each wf onto a statement form A' in such a way that if A is a theorem of $\mathfrak{T}$ then A' is a tautology. Then a wf B can be constructed such that B' is not a tautology; it follows that B is not a theorem. We state this result as the following theorem.

THEOREM 3.7.3. A first-order predicate calculus is consistent.

Sometimes the notion of completeness, in the sense of one or more of the definitions given in §3.3, may be treated in metamathematics. For example, Theorem 3.5.3, which asserts the completeness of the statement calculus in a positive sense (as this was explained in §3.3), belongs to metamathematics. On the other hand, Gödel's completeness theorem for a predicate calculus is outside the realm of metamathematics. The statement calculus is also complete in a sense which exhibits a negative approach to a sufficiency of theorems. The next result, which belongs to metamathematics, is of this sort.

THEOREM 3.7.4. If A is any wf of the statement calculus, then either it is a theorem or else an inconsistent theory results by adding as additional axioms all wfs resulting from A by substituting arbitrary wfs for its statement letters.

Proof. Let A be a wf which is not a theorem, and let us augment the axiom schemas of the statement calculus with all wfs resulting from A by substituting arbitrary wfs for its statement letters. Since A is not a theorem, it is not a tautology. Therefore, it takes the value F for some row of its truth table. Referring to one such row, we construct a wf as follows. In A substitute $A \vee \sim A$ for the statement letters

---

* A detailed proof is given in Margaris (1967), Section 25.

which are T and substitute $A \wedge \sim A$ for those statement letters which are F. The resulting axiom, B, will always take the value F. Then $\sim B$ is a tautology, and hence a theorem. Thus, both B and $\sim B$ are theorems.

Decidability questions are metamathematical in nature.* We recall that for formal axiomatic theories the notion of proof is effective, that is, there exists an effective procedure for deciding whether a finite sequence of wfs is a proof. In contrast to the question of whether a given sequence of wfs is a proof, which requires merely the examination of a displayed finite object, that of whether a given wf is a theorem requires looking elsewhere than within the given object for an answer. Further, the definition of proof sets no bounds on the length of a proof, and to examine all possible proofs without a bound on their lengths is not a procedure which yields an answer to the question in a finite number of steps if the wf is not a theorem. Thus, the decision problem for provability has special significance for formal theories. Accordingly, it is often called *the* decision problem for a theory. Our earlier definition of a decidable theory, as one for which the theorems constitute a decidable set, takes this special significance into account. We recall that in §3.5 it was noted that the statement calculus is a decidable formal axiomatic theory.

The first significant theorems regarding undecidability were proved in 1936 by the American logician Alonzo Church. We shall discuss two of his results, both of which are usually identified simply as **Church's theorem.** One of these asserts that both the predicate calculus containing all predicate letters, all function letters, and all individual constants and the predicate calculus containing all predicate letters but no function letters or individual constants are undecidable theories. This is an "impossibility theorem" similar in nature to the theorem which asserts that, using just ruler and compass, there is no method for trisecting an arbitrary angle. The analogue, for Church's theorem, of specifying the admissible methods (that is, constructions by ruler and compass) is specifying methods for determining what sets are decidable. At the intuitive level the answer is given in terms of an effective procedure, and at the rigorous level the answer is given in terms of a concept that is identified by the word "recursive." The definition of "recursively decidable" appears in the Appendix to this chapter. The absence of a

---

* Detailed accounts of the results touched on in the remainder of this section may be found in Mendelson (1964).

definition is of no handicap here. It suffices to note that there is no way to prove that effective decidability coincides with recursive decidability, because the former is an intuitive concept. Rather, one can accept or reject the hypothesis that recursive decidability captures the essence of the notion of effective decidability. The hypothesis that recursive decidability corresponds to effective decidability is known as **Church's thesis.** In the above statement of Church's theorem, "undecidable" should be modified by the word "recursively," for that is what Church *proved*. One who accepts Church's thesis, and most do on the basis of the evidence that is available, is prepared to substitute "effectively" for "recursively."

If we accept his thesis, Church's theorem implies that there is no effective method for determining whether any given wf is logically valid. It also implies, in conjunction with results that we shall not state, that there is no decision procedure for provability or validity of wfs containing only two-place predicate letters. (In contrast, the predicate calculus having just one-place predicate letters and neither function letters nor individual constants is effectively decidable.) The other theorem that bears Church's name is closely related to that given above. To state it in a general form requires a definition. A first-order theory $\mathfrak{T}'$ having the same alphabet as the first-order $\mathfrak{T}$ is called an **extension** of $\mathfrak{T}$ if every theorem of $\mathfrak{T}$ is a theorem of $\mathfrak{T}'$. The promised theorem then reads as follows: First-order arithmetic (that is, the theory N defined in Example 3.6.3) and all consistent extensions of it are undecidable.* As in the earlier theorem "undecidable" should be modified by "recursively" or "effectively." Since 1936 a variety of theories, including group theory and lattice theory, have been shown to be undecidable. In addition, a variety of interesting theories have been shown to be decidable.

We have saved for last in this brief survey of metamathematics the most outstanding results. They are fundamental since they have direct impact on the questions of the consistency and completeness of classical mathematics—the very problems which led Hilbert to formulate his program. These results were proved by Gödel in 1931. To state them in a general form,† let us call a first-order theory $\mathfrak{T}$ (in which equality is definable) **sufficiently rich** if at least first-order arithmetic can be developed within $\mathfrak{T}$. This means that certain function letters of $\mathfrak{T}$ can be

---

* Actually, Church proved that N is undecidable, and J. B. Rosser extended this result to its consistent extensions.

† This general form assumes Church's thesis.

identified with those introduced for the theory N and the proper axioms of N can be proved as theorems of $\mathfrak{T}$. The usual axiomatic theories of sets are sufficiently rich, for example. A generalized form of Gödel's first theorem states that if the proper axioms of a first-order theory are required to be a decidable set—that is, the theory is required to be axiomatic (see §3.4)—then it is impossible to construct a sufficiently rich, consistent first-order theory in which all true statements expressible within the theory are provable as theorems. Indeed, in a sufficiently rich, consistent, and axiomatic first-order theory, one can construct a closed wf such that neither it nor its negation is provable; that is, each such theory contains a so-called **undecidable statement.** Thus, in particular, there is a statement S of N such that neither S nor $\sim S$ is a theorem. Hence, N is negation incomplete. Since these wfs are closed, one is true and the other is false when interpreted as number-theoretic statements. Since neither is provable, there is a true statement of number theory which is unprovable.

One might hope that the defect of incompleteness of N could be remedied by adjoining, as an axiom, either S or $\sim S$, where S is an undecidable statement, or in other words, that a suitable extension of N would be complete. This avenue leads nowhere, for according to the theorem, if the extension of N is both consistent and axiomatic, then it too contains an undecidable statement.

Gödel's second theorem, which can be derived from his first, asserts that the consistency of a sufficiently rich, consistent, first-order theory $\mathfrak{T}$ cannot be proved by methods which can be expressed or reflected in $\mathfrak{T}$. For N itself this means that, if consistent, we cannot prove the consistency of N within N. Instead, we must use the methods that go beyond those expressible within N, provided N is consistent. Gödel suggested the possibility of a proof which, although it cannot be reflected in N, would still be finitary. So far no positive results have been obtained in this direction. Indeed, the only proof of the consistency of N (by Gentzen) uses transfinite induction and so is therefore not finitary. For a complete understanding of the foregoing, precise meanings must be assigned to the expressions "expressible within N" and "reflected in N." This becomes possible once the arithmetization of metamathematics by means of Gödel's numbering is understood. This topic is discussed in Stoll, *Set Theory and Logic.*

APPENDIX

Recursive Functions and Church's Thesis

THIS APPENDIX SUPPLIES the concepts which are necessary to lend precision to some of the notions described and theorems stated in §§ 3.4 through 3.7.

Our concern is with number-theoretic functions, that is, functions on $\mathbb{N}^r$ into $\mathbb{N}$, $r = 1, 2, \ldots$. If there is an algorithm (in the sense of the intuitive description given in §3.4) for computing values of such a function, we shall say that it is **effectively computable.** Our first objective is to describe a class of number-theoretic functions whose members we feel justified in classifying as effectively computable. The class is generated by specifying certain members (called initial functions) and rules for obtaining further functions from functions known to fall in the class.

The **initial functions** are the zero function $Z : \mathbb{N} \to \mathbb{N}$ such that $Z(x) = 0$ for all x, the successor function $S : \mathbb{N} \to \mathbb{N}$ such that $S(x) = x + 1$ for all x, and the projection functions $U_i^n : \mathbb{N}^n \to \mathbb{N}$ where, for each $1 \le i \le n$ and $n = 1, 2, \ldots$, $U_i^n (x_1, x_2, \ldots, x_n) = x_i$. One rule for obtaining further functions is that of composition: Given n functions $h_1, \ldots, h_n$, each being of m arguments, and given a function g of n arguments, a function f of m arguments is defined by

$$f(x_1, \ldots, x_m) = g(h_1(x_1, \ldots, x_m), \ldots, h_n(x_1, \ldots, x_m)).$$

Another rule is that of recursion. This associates with a function g of $n + 2$ arguments and a function h of n arguments a function f of $n + 1$ arguments defined by

$$f(x_1, \ldots, x_n, 0) = h(x_1, \ldots, x_n),$$
$$f(x_1, \ldots, x_n, y + 1) = g(x_1, \ldots, x_n, y, f(x_1, \ldots, x_n)).$$

Here we allow the value 0 for n, in which case

$$f(0) = c \text{ (a fixed natural number)},$$
$$f(y + 1) = g(y, f(y)).$$

In keeping with the definition given in §1.12 (immediately after Theorem 1.12.2), we shall say that f is obtained from g and h (or, in the case $n = 0$, from g alone) by recursion.

The remaining rule is that of minimalization. This associates a function f of n arguments with a given function h of $n + 1$ arguments by the rule

$$f(x_1, \ldots, x_n) = \min y \, (h(x_1, \ldots, x_n, y) = 0),$$

where "min y" means "the smallest y such that." Usage of this rule is restricted (for our purposes, at least) to those cases in which a minimum exists for every set of values of the parameters x_i.

We now define a **recursive function** to be any (number-theoretic) function that either is an initial function or can be obtained from initial functions by a finite number of applications of the three rules. A **primitive recursive function** is a recursive function that can be obtained from initial functions without any applications of minimalization. There are recursive functions which are not primitive recursive.* A wide variety of number-theoretic functions are primitive recursive† and, hence, recursive. We feel justified in judging recursive functions to be computable—certainly that is true of the initial functions, and the operations of composition, recursion, and minimalization produced computable functions from computable functions. What about the converse? It is at this point that **Church's thesis** enters. Church's thesis asserts that recursiveness is equivalent to effective computability. Thus it serves to give precision to the intuitive notion of a computable function, that is, a number-theoretic function whose values are computed by preassigned rules. Because of the vagueness of the intuitive notions of effectively computable functions, it is impossible to *prove* Church's thesis, although considerable evidence in support of it has grown out of attempts to analyze the notion of algorithm.‡

Accepting Church's thesis as we do, precision can be achieved in some of the notions and topics discussed in Chapter 3, by means of the arithmetization of formal theories, which is based on an idea devised by Gödel. We illustrate the method for an arbitrary first-order theory $\mathfrak{T}$. The first step is to associate with each symbol s of $\mathfrak{T}$ a positive integer $g(s)$, called the **Gödel number** of s, so that different symbols have different Gödel numbers. For example, the following is such an assignment:

$$g(\sim) = 3, \quad g(\to) = 5, \quad g(() = 7, \quad g()) = 9, \quad g(,) = 11,$$
$$g(x_j) = 5 + 8j \quad \text{for } j = 1, 2, \ldots$$
$$g(a_j) = 7 + 8j \quad \text{for } j = 1, 2, \ldots$$

---

* See Mendelson (1964), p. 250, for an example.
† See *ibid.*, Chapter 3.
‡ A thorough discussion of arguments for Church's thesis may be found in Kleene (1952), Chapters 12 and 13.

$$g(P_j^n) = 9 + 8(2^n 3^j) \quad \text{for } n \geq 0, j \geq 1$$
$$g(f_j^n) = 11 + 8(2^n 3^j) \quad \text{for } n, j \geq 1.$$

Next, this assignment of positive (and, in this case, odd) integers to symbols is extended to an assignment of positive integers to expressions: to the expression $s_1 s_2 \cdots s_r$ is assigned as Gödel number

$$g(s_1 s_2 \cdots s_n) = 2^{g(s_1)} \cdot 3^{g(s_2)} \cdot \cdots \cdot p_r^{g(s_r)}$$

where p_i is the ith prime and $p_1 = 2$. Different expressions have different Gödel numbers by virtue of the uniqueness of the factorization of integers into a product of powers of primes (see Example 1.12.5). Moreover, expressions have different Gödel numbers than symbols do, since the Gödel numbers of expressions are even and those of symbols are odd.* We assign a finite sequence $e_1, e_2, \ldots, e_r$ of expressions e_i a Gödel number, by setting

$$g(e_1, e_2, \ldots, e_r) = 2^{g(e_1)} \cdot 3^{g(e_2)} \cdots \cdot p_r^{g(e_r)}.$$

Different sequences of expressions have different Gödel numbers. Since the Gödel number of a sequence of expressions is even and the exponent of 2 in its factorization into primes is also even, it differs from Gödel numbers of symbols and expressions. Piecing together the foregoing, it follows that g is a one-to-one function on the union of the set of symbols of $\mathfrak{T}$, the set of all expressions of $\mathfrak{T}$, and the set of all finites equences of expressions of $\mathfrak{T}$ into the set of positive integers. The range of g is a proper subset of this set; for example, 6 is not a Gödel number.

In a similar fashion, once the alphabet of *any* formal theory has been coded by assigning positive integers to its members (with distinct symbols being assigned different integers), the function so defined can be extended to a one-to-one function on the alphabet of the theory, its expressions, and all finite sequences of expressions into the set of positive integers. Such a map is called a **Gödel numbering** of the theory.

According to a definition given in §3.4, a set of natural numbers is effectively decidable iff there exists an algorithm for deciding whether any natural number is an element of that set. We shall identify these sets with so-called **recursive sets,** that is, sets of natural numbers whose characteristic functions are recursive. It can be shown that this identification is equivalent to Church's thesis—the identification of effectively computable functions with recursive functions. With the goal of providing some "feeling" for recursive sets, we make a definition and prove

* A single symbol, considered as an expression, has a different number from its number as a symbol. This fact need not cause any confusion.

a theorem. A set S of natural numbers is called **recursively enumerable** iff it is empty or is the range of a recursive function. If S is recursively enumerable, a function whose range is S is said to generate S. Accepting Church's thesis, this means that intuitively a recursively enumerable set is a collection of natural numbers which is generated by some algorithm. The relationship between the notions of recursiveness and recursive enumerability is given next.

THEOREM.   A set S is recursive iff both S and its complement $\bar{S}$ (in $\mathbb{N}$) are recursively enumerable.

Proof.   Assume that S is recursive. If S is empty it is recursively enumerable by definition. Next, suppose S is finite and nonempty, say, $S = \{n_0, n_1, \ldots, n_r\}$. Define g by

$$g(x) = \begin{cases} n_x & \text{for } x \leq r \\ n_r & \text{for } x > r. \end{cases}$$

Finally, suppose S is infinite. Let f be its characteristic function. Define g by

$$g(0) = \min y[f(y) = 1],$$

$$g(n + 1) = \min y[(sg(y \div g(n)) \cdot f(y)) = 1].^*$$

Hence, in every case, S is recursively enumerable. Since S recursive implies $\bar{S}$ is recursive (and, hence, recursively enumerable by the above), it follows that S recursive implies that both S and $\bar{S}$ are recursively enumerable.

For the converse, assume that S and $\bar{S}$ are recursively enumerable. If either S or $\bar{S}$ is empty, S is immediately recursive. If neither is empty, then S is the range of a recursive function f and $\bar{S}$ is the range of another such function g. Now f and g give an effective procedure for testing for membership in S. To test a given x, examine in turn $f(0), g(0), f(1), g(1), f(2), \ldots$. Since $S \cup \bar{S} = \mathbb{N}$, x must appear as a value of f or g. If x appears as a value of f, then $x \in S$; if x appears as a value of g, then $x \notin S$.

The foregoing is a sufficient explanation for a proof by Church's thesis.

Returning to the definition of a formal theory $\mathfrak{T}$ in §3.4, we may recast the three requirements that accompany that definition in terms of a Gödel numbering of $\mathfrak{T}$. They become

---

* The function sg is defined as follows: $sg(n) = 0$ if $n = 0$ and $sg(n) = 1$ if $n \neq 0$. The function $\div$ is defined by $r \div s = r - s$ if $r \geq s$ and $r \div s = 0$ if $r < s$. Both functions can be shown to be primitive recursive.

(i)′ The set of Gödel numbers of the alphabet of $\mathfrak{T}$ is recursive.

(ii)′ The set of Gödel numbers of the wfs of $\mathfrak{T}$ is recursive.

(iii)′ The set of Gödel numbers of each rule of inference is recursive.

Also, we may now say that $\mathfrak{T}$ is an axiomatic theory if the set of Gödel numbers of its axioms is recursive. Further, $\mathfrak{T}$ is recursively decidable if the set of Gödel numbers of its theorems is recursive and recursively undecidable if the set of Gödel numbers of its theorems is not recursive.

CHAPTER FOUR

Boolean Algebras

THE THEORY OF Boolean algebras has historical as well as present-day practical importance. For the beginner its exposition should prove a serviceable vehicle for assimilating many of the concepts discussed in general terms in Chapter 3. Moreover, it illustrates the important type of axiomatic theory known as an "algebraic theory." The theory of Boolean algebras is, on one hand, relatively simple and, on the other hand, exceedingly rich in structure. Thus, its detailed study serves in some respects as an excellent introduction to techniques which one may employ in the development of a specific axiomatic theory. The only possible shortcoming is that the ease with which it may be put into a relatively completed form is somewhat misleading, since, for axiomatic theories in general, this task is not so easy.

This chapter presents first a natural formulation of the theory, then a formulation which is commonly regarded as being more elegant. This second formulation is used to develop the next topic, the representation theory for Boolean algebras in terms of algebras of sets. In conclusion, the notion of a free Boolean algebra is presented.

§4.1. A Definition of a Boolean Algebra

By an **algebra of sets** based on U we shall mean a nonempty collection $\mathcal{C}$ of subsets of the nonempty set U such that if A, $B \in \mathcal{C}$, then $A \cup B$, $A \cap B \in \mathcal{C}$, and if $A \in \mathcal{C}$, then $\overline{A} \in \mathcal{C}$. For example, the power set of U, $\wp(U)$, is an algebra of sets. However, certain proper subsets of $\wp(U)$ may be an algebra of sets (see Exercise 4.2.6). If $\mathfrak{a}$ is an algebra of sets based on U, then $U \in \mathfrak{a}$ (since if $A \in \mathfrak{a}$, then $U = A \cup \overline{A} \in \mathfrak{a}$) and $\varnothing \in \mathfrak{a}$ (since if $A \in \mathfrak{a}$, then $\varnothing = A \cap \overline{A} \in \mathfrak{a}$). Further, Theorem 1.5.1 may be interpreted as a list of properties of an algebra of sets. That this is a fundamental list of properties is suggested by the variety of other properties (for example, those in Theorem 1.5.2) which may be deduced solely from them. As formulated below, the theory of Boolean algebras is the axiomatized version of algebras of sets when viewed as structures having the properties appearing in Theorem 1.5.1. The omission of the associative laws for the two binary operations in the list of axioms is deliberate; they will be derived as theorems.

A **Boolean algebra** is a structure $\langle B, \cup, \cap, ', 0, 1 \rangle$ where B is a set, $\cup$ is a binary operation (called **union** or **join**) in B, $\cap$ is a binary operation (called **intersection** or **meet**) in B, $'$ is a unary operation (called **complementation**) in B, 0 and 1 are elements of B, and the following nine axioms are satisfied.

For all a, $b \in B$,

  (i) $a \cup b = b \cup a$,   (i)' $a \cap b = b \cap a$.

For all a, b, $c \in B$,

  (ii) $a \cup (b \cap c) = (a \cup b) \cap (a \cup c)$,

                        (ii') $a \cap (b \cup c) = (a \cap b) \cup (a \cap c)$.

For all $a \in B$,

  (iii) $a \cup 0 = a$,   (iii)' $a \cap 1 = a$.

For all $a \in B$,

  (iv) $a \cup a' = 1$,   (iv)' $a \cap a' = 0$.

The elements 0 and 1 are distinct; that is,

  (v) $0 \neq 1$.

The consistency of the theory that we have just formulated can be established by choosing for B the power set of a nonempty, finite set U, taking $\cup$ and $\cap$ as set-union and set-intersection, respectively, $'$ as complementation relative to U, and, finally, choosing 0 and 1 as $\varnothing$ and U, respectively. The uniqueness of the elements 0 and 1 is established in Theorem 4.2.1. These uniquely determined elements are called the **zero element** and **unit element,** respectively, of a Boolean algebra. It was in anticipation of this uniqueness and terminology that the symbols "0" and "1" were used in the axioms. We might have postulated their uniqueness; however, we would then be obligated to *prove* uniqueness as part of any verification that an alleged Boolean algebra is truly just that. If a is an element of a Boolean algebra, its image a' under the unary operation assigned to the algebra is called the **complement** of a.

The above formulation of the theory of Boolean algebras is essentially that given in 1904 by the American mathematician E. V. Huntington (1875–1945) who devoted himself to devising independent sets of axioms for a variety of algebraic theories. The independence of the above set of nine axioms is discussed in the following exercises.

Exercises

The status of the above axiom system for Boolean algebras is this: Each of the seven axioms (i), (i)′, (ii), (ii)′, (iv), (iv)′, and (v) is independent and neither (iii) nor (iii)′ are independent. Equivalently, the set of nine axioms with either (iii) or (iii)′ deleted is an independent set of axioms. The independence of (i), (i)′, and so on, can be demonstrated by models of the set of all axioms except one of the seven proclaimed independent ones. For a set B having just a few elements, an operation in B can be defined by a table. If the operation is binary, the table is a square array whose rows and columns are numbered with the elements of B and such that at the intersection of the ath row and the bth column the composite of a and b appears. If the operation is unary, a table having just two columns is adequate. For example, Table 4.1 defines such operations in the set $B = \{0,1\}$.

Table 4.1.

$\cup$	0	1
0	0	1
1	1	0

$\cap$	0	1
0	0	0
1	0	1

′	
0	1
1	0

4.1.1. Show that $\langle\{0,1\},\cup,\cap,',0,1\rangle$, where the operations in $\{0,1\}$ are those defined above, establishes the independence of axiom (ii). Next show that (ii)′ is independent, using the same structure but with operations defined as in Table 4.2.

Table 4.2.

$\cup$	0	1
0	0	1
1	1	1

$\cap$	0	1
0	1	0
1	0	1

′	
0	1
1	0

4.1.2. Construct five other structures which demonstrate the independence of each of (i), (i)', (iv), (iv)', and (v).

4.1.3. Derive axiom (iii) from the others.

4.1.4. Derive axiom (iii)' from the others.

§4.2. Some Basic Properties of a Boolean Algebra

The properties of a Boolean algebra which are derived in this section are the abstract versions of the results obtained in §1.5 following Theorem 1.5.1 for an algebra of sets. The only essential difference is that now the axioms of a Boolean algebra are used in place of Theorem 1.5.1.

We begin with the remark that once $\cup$ and $\cap$ are shown to be associative operations, there follows the general associative law and the general commutative law for each operation, as well as the general distributive law for each operation with respect to the other. Paralleling the presentation in §1.5, we define next the **dual** of a Boolean statement, that is, a closed formula of the language accompanying the theory, as the statement which results from the original upon the replacement of $\cup$ by $\cap$ and $\cap$ by $\cup$, 1 by 0 and 0 by 1. It is evident that the dual of each axiom is an axiom, with (v) being its own dual. Hence, if T is any theorem derivable from (i), (i)', . . . , (v), then the dual of T is derivable from the set of axioms, the duals of the steps in the proof of T providing a proof of the dual. This is the (proof-theoretic version of the) **principle of duality** for the theory of Boolean algebras*; it yields a companion theorem for each theorem that has been proved unless that theorem is its own dual. The reader can construct an illustration of the principle by writing down the dual of each step in the proof below of the identity $a \cup a = a$ [part (x) of Theorem 4.2.1] to obtain a proof of $a \cap a = a$. The collection of results which form this theorem is the Boolean algebra version of Theorem 1.5.2 plus the associative laws for $\cup$ and $\cap$.

THEOREM 4.2.1. In each Boolean algebra $\langle B, \cup, \cap, ', 0, 1 \rangle$ the following hold.

  (vi) The elements 0 and 1 are unique.

  (vii) Each element has a unique complement; that is, if $a \cup b = 1$ and $a \cap b = 0$, then $b = a$.

  (viii) For each element a, $(a')' = a$.

  (ix) $0' = 1$ and $1' = 0$.

---

* A more sophisticated form of the principle asserts that a Boolean statement is true for a Boolean algebra iff its dual is true for that algebra.

(x) For each element a, $a \cup a = a$ and $a \cap a = a$.

(xi) For each element a, $a \cup 1 = 1$ and $a \cap 0 = 0$.

(xii) For all a and b, $a \cup (a \cap b) = a$ and $a \cap (a \cup b) = a$.

(xiii) For all a and b, $(a \cup b)' = a' \cap b'$ and $(a \cap b)' = a' \cup b'$.

(xiv) For all a, b, and c, $a \cup (b \cup c) = (a \cup b) \cup c$ and $a \cap (b \cap c) = (a \cap b) \cap c$.

Proof. For (vi) assume that along with 0, the element 0_1 has the property that $a \cup 0_1 = a$ for all a. Then $0_1 \cup 0 = 0_1$ and $0 \cup 0_1 = 0$. By axiom (i), $0_1 \cup 0 = 0 \cup 0_1$ and, hence, $0_1 = 0$. The uniqueness of 1 follows by the principle of duality.

For (vii) assume that $a \cup b = 1$ and $a \cap b = 0$ for elements a and b. Then

$$
\begin{aligned}
b &= b \cup 0 & &\text{by (iii)};\\
&= b \cup (a \cap a') & &\text{by (iv)}';\\
&= (b \cup a) \cap (b \cup a') & &\text{by (ii)};\\
&= (a \cup b) \cap (a' \cup b) & &\text{by (i)};\\
&= 1 \cap (a' \cup b) & &\text{by hypothesis};\\
&= (a \cup a') \cap (a' \cup b) & &\text{by (iv)};\\
&= (a' \cup a) \cap (a' \cup b) & &\text{by (i)};\\
&= a' \cup (a \cap b) & &\text{by (ii)};\\
&= a' \cup 0 & &\text{by hypothesis};\\
&= a' & &\text{by (iii)}.
\end{aligned}
$$

For (viii), by definition of the complement of a, $a \cup a' = 1$ and $a \cap a' = 0$. Hence, by (i), $a' \cup a = 1$ and $a' \cap a = 0$. That is, $(a')' = a$, by (vii).

The proof of (ix) is left as an exercise.

The proof of (x) is the following computation:

$$
\begin{aligned}
a \cup a &= (a \cup a) \cap 1 & &\text{by (iii)}';\\
&= (a \cup a) \cap (a \cup a') & &\text{by (iv)};\\
&= a \cup (a \cap a') & &\text{by (ii)};\\
&= a \cup 0 & &\text{by (iv)}';\\
&= a & &\text{by (iii)}.
\end{aligned}
$$

Proofs of (xi) to (xiii) are left as exercises.

The proof of the associative law for $\cup$, the first assertion in (xiv), proceeds as follows. Let

$$x = a \cup (b \cup c) \quad \text{and} \quad y = (a \cup b) \cup c.$$

Then $a \cap x = a$ and $a \cap y = a$ using (xii). Further, $a' \cap x =$

$a' \cap (a \cup (b \cup c)) = a' \cap (b \cup c)$ and, similarly, $a' \cap y = a' \cap (b \cup c)$. Hence

$$a \cap x = a \cap y \qquad \text{and} \qquad a' \cap x = a' \cap y.$$

It follows, in turn, that

$$(a \cap x) \cup (a' \cap x) = (a \cap y) \cup (a' \cap y),$$
$$(x \cap a) \cup (x \cap a') = (y \cap a) \cup (y \cap a'),$$
$$x \cap (a \cup a') = y \cap (a \cup a'),$$
$$x \cap 1 = y \cap 1,$$
$$x = y.$$

It is possible to introduce into the set B of a Boolean algebra $\langle B, \cup, \cap, ', 0, 1 \rangle$ a partial-ordering relation which resembles that of set inclusion. The characterization of inclusion in Theorem 1.5.3 in terms of set intersection is the origin of the following definition. If $\langle B, \cup, \cap, ', 0, 1 \rangle$ is a Boolean algebra, and a, b are elements of B such that $a \cap b = a$ (or, equivalently, $a \cup b = b$), then we write

$$a \le b \qquad \text{or} \qquad b \ge a.$$

From these definitions it follows immediately that the principle of duality may be extended to statements where $\le$ or $\ge$ are present by adding the provision that $\le$ should be replaced by $\ge$ and conversely. Several properties of $\le$ are stated in Exercise 4.2.4. Others are gathered in the next theorem.

THEOREM 4.2.2. If $\langle B, \cup, \cap, ', 0, 1 \rangle$ is a Boolean algebra, then $\langle B, \le \rangle$ is a partially ordered set with greatest element (namely, 1) and least element (namely, 0). Moreover, each pair $\{a, b\}$ of elements has a least upper bound (namely, $a \cup b$) and a greatest lower bound (namely, $a \cap b$).

Proof. The proof is straightforward and is left as an exercise.

Exercises

4.2.1. Referring to Theorems 1.5.2 and 4.2.1, it is obvious that (viii) to (xiii) of Theorem 4.2.1 are the abstract versions of 8, 8′ to 13, 13′ of Theorem 1.5.2. Show that (vi) and (vii) of Theorem 4.2.1 are the abstractions of 6, 6′ and 7, 7′, respectively, of Theorem 1.5.2.

4.2.2. Supply proofs for parts (ix), (xi), (xii), and (xiii) of Theorem 4.2.1.

4.2.3. Let $\langle B, \cup, \cap, ', 0, 1 \rangle$ be a Boolean algebra and A be a nonempty subset of B that is closed under the operations $\cup$, $\cap$, $'$ (thus, for elements a and b

in A, $a \cup b$, $a \cap b$, and a' are in A). Show that 0 and 1 are elements of A and that $\langle A, \cup_A, \cap_A, '_A, 0, 1 \rangle$, where $\cup_A$, $\cap_A$, $'_A$ are the restrictions of $\cup$, $\cap$, $'$, respectively, to A, is a Boolean algebra.

4.2.4. Establish each of the following as a theorem for Boolean algebras.

   (a) $a \cup b = b$ iff $a \cap b = a$.

   (b) $a \leq b$ iff $a \cap b' = 0$ iff $a' \cup b = 1$.

   (c) $a \leq b$ iff $b' \leq a'$.

   (d) For given x and y, $x = y$ iff $0 = (x \cap y') \cup (y \cap x')$.

4.2.5. Prove Theorem 4.2.2.

4.2.6. Show that the results stated in Exercise 1.11.18 together with the principle of duality provide a proof of Theorem 4.2.2.

4.2.7. Let B be the set whose elements are the natural numbers 0 and 1. For $a, b \in B$, define $a \cup b = a + b - a \cdot b$, $a \cap b = a \cdot b$, and $a' = 1 - a$. Show that $\langle B, \cup, \cap, ', 0, 1 \rangle$ is a Boolean algebra.

4.2.8. Construct a Boolean algebra whose domain is the set having the truth values T and F as its members. Describe the relationship between this algebra and that in the preceding exercise.

4.2.9. Let $\mathfrak{A}$ be the collection of all subsets A of $\mathbb{Z}^+$ such that either A or $\overline{A}$ is finite. Show that $\langle \mathfrak{A}, \cup, \cap, {}^-, \varnothing, \mathbb{Z}^+ \rangle$ where the operations are the familiar set-theoretical union, intersection, and complementation, is a Boolean algebra.

Remark

The remaining problems in this section are concerned with a type of generalization of a Boolean algebra called a **lattice**. A lattice is a triple $\langle X, \cup, \cap \rangle$, where X is a nonempty set, $\cup$ and $\cap$ are binary operations in X (read "union" and "intersection," respectively), and the following axioms are satisfied. For all a, b, $c \in X$,

L_1. $a \cup (b \cup c) = (a \cup b) \cup c$; L_1'. $a \cap (b \cap c) = (a \cap b) \cap c$;

L_2. $a \cup b = b \cup a$; L_2'. $a \cap b = b \cap a$;

L_3. $(a \cup b) \cap a = a$; L_3'. $(a \cap b) \cup a = a$.

4.2.10. Verify the following properties of a lattice.

   (a) For all a, $a \cup a = a$ and $a \cap a = a$.

   (b) For all a, b, the relations $a \cup b = a$ and $a \cap b = b$ are equivalent.

   (c) For all a, b, the relations $a \cap b = a$ and $a \cup b = b$ are equivalent.

4.2.11. Let $\langle X, \leq \rangle$ be a partially ordered set such that each pair of elements has a least upper bound and a greatest lower bound in X. Thus, if we set $a \cup b = \text{lub } \{a,b\}$ and $a \cap b = \text{glb } \{a,b\}$, then $\cup$ and $\cap$ are operations in X. Prove that $\langle X, \cup, \cap \rangle$ is a lattice. Next, prove that, conversely, if in a lattice $\langle X, \cup, \cap \rangle$ we define the relation $\leq$ by $a \leq b$ iff $a \cap b = a$, then $\langle X, \leq \rangle$ is a partially ordered set such that each pair of elements has a least upper bound (namely, $a \cup b$) and a greatest lower bound (namely, $a \cap b$).

Remark

This result gives, in effect, a second formulation of the axiomatic theory called lattice theory. Thus, one may think of a lattice in either way. If the formulation is in terms of $\leq$, then, by $\cup$ and $\cap$, one understands the operations in Exercise 4.2.11. If the formulation is in terms of $\cup$ and $\cap$, then, by $\leq$, one understands the ordering relation defined, again, in Exercise 4.2.11.

4.2.12. Let $\langle X,\cup,\cap\rangle$ and $\langle X',\cup',\cap'\rangle$ be lattices. Regarded as partially ordered sets, $\langle X,\leq\rangle$ and $\langle X',\leq'\rangle$, they are isomorphic (see §1.11) iff there exists a one-to-one correspondence f between X and X' such that both f and f^{-1} are order-preserving. What is the condition for isomorphism of these lattices in terms of the binary operations of union and intersection?

4.2.13. Show that there are exactly five nonisomorphic lattices of fewer than five elements and that there are exactly five nonisomorphic lattices of five elements. (Hint: For this problem it is more convenient to think of a lattice as a partially ordered set.)

4.2.14. Show that if a lattice satisfies either of the identities

$$a \cap (b \cup c) = (a \cap b) \cup (a \cap c) \quad \text{all } a, b, c,$$
$$a \cup (b \cap c) = (a \cup b) \cap (a \cup c) \quad \text{all } a, b, c,$$

then it satisfies the other. *Note.* A lattice which satisfies one, hence, both of the above identities is called **distributive.**

4.2.15. A lattice $\langle X,\cup,\cap\rangle$ is called **complemented** if it has a least element, 0, and a greatest element, 1, and for each element a of X there is at least one element b in X such that $a \cap b = 0$ and $a \cup b = 1$. Show that if $\langle X,\cup,\cap\rangle$ is both distributive and complemented, then each element has a unique complement.

4.2.16. Show that a lattice is a Boolean algebra iff it is both complemented and distributive.

Remark

It is possible to discuss at this point the principle of duality for Boolean algebras in a more systematic way than we have done. What follows illustrates the way in which this principle applies to algebra and projective geometry. Basic to our presentation is our point of view that a Boolean algebra is a type of partially ordered set. Thus we begin by recalling that if $P = \langle X,\geq\rangle$ is a partially ordered set, and we define the binary relation $\leq^d$ in X by $a \leq^d b$ iff $b \leq a$, then $P^d = \langle X,\leq^d\rangle$ is a partially ordered set which is called the **dual** of P. For a given statement S formulated in the language of P, we can form the **dual statement,** S^d, by replacing each of the relations $\leq$ and $\geq$ by the other wherever they occur in S. As an illustration, the dual of "each nonempty subset of X contains a maximal element" is "each nonempty subset of X contains a minimal element." A statement of P is called **self-dual** iff whenever it is true for P it is also true for P^d. For example, the statement "P is a chain" is self-dual, but the statement "each nonempty subset of X contains a maximal element" is not. It follows that if a statement of P is self-dual and is true for P, then the dual of this statement is true for P. Thus, if a statement is derived as a consequence of certain self-dual statements of P, then S^d is true for P. This is the **principle of duality** for the theory of partially ordered sets.

If the partially ordered set $L = \langle X, \leq \rangle$ is a lattice, then so is its dual $L^d = \langle X, \geq \rangle$; the identity map on X takes lub $\{a,b\}$ in L into glb $\{a,b\}$ in L^d, and vice versa. Hence, every property of a lattice has a dual property which is characterized by interchanging $\cup$ and $\cap$. Since properties $L_1, L_2, \ldots, L_3'$ characterize these operations and each is self-dual, the duality principle for lattices reads: if T is a theorem of L, then its dual T^d, obtained from T by interchanging $\leq$ and $\geq$, and $\cup$ and $\cap$, is a theorem of L. If L has a zero element 0 and a unit element 1, then 0 and 1 are dual to each other, that is, $0 \leq a \leq 1$ and $1 \geq a \geq 0$ for all $a \in X$. The dual of a statement involving 0 or 1 is obtained by interchanging 0 and 1 (as well as $\leq$ and $\geq$, and $\cup$ and $\cap$).

Next, suppose that the partially ordered set $\langle B, \leq \rangle$ is a Boolean algebra, that is, a complemented, distributive lattice. The existence of a unique complement for each element $a \in B$ is a self-dual property of a lattice. Further, according to Exercise 4.2.14, that of being distributive is self-dual. Hence, the duality principle for lattices, as stated above, holds for Boolean algebras, and the dual lattice of a Boolean algebra is also a Boolean algebra. In fact, the reader should prove that the mapping which assigns to each element of B its complement is an isomorphism between $\langle B, \leq \rangle$ and its dual. Notice, finally, that we have established the principle of duality for Boolean algebras as stated in a footnote at the beginning of this section: a Boolean statement is true for a Boolean algebra iff its dual is true for the algebra.

§4.3. Another Formulation of the Theory

The formulation which we have given of the theory of Boolean algebras has much to recommend it. The primitive terms are few, and the simplicity and symmetry of the axioms lend aesthetic appeal. Moreover, if (iii) or (iii)$'$ is omitted, the resulting set is independent. Finally, the formulation clearly reflects the type of system that motivated it. However, it is always a challenge to see if a formulation can be pared down in one or more respects. For Boolean algebras this challenge has been successfully met by a great variety of formulations. We shall describe one that has become quite popular. It achieves for arbitrary Boolean algebras the analogue of the familiar fact for an algebra of sets that either of the operations of union and intersection can be eliminated in terms of the other together with complementation (for example, $A \cup B = (\overline{\overline{A} \cap \overline{B}})$).

If $\langle B, \cup, \cap, ', 0, 1 \rangle$ is a Boolean algebra, then B is a set with at least two distinct members. Moreover, the binary operation $\cap$ and the unary operation $'$ have the following properties.

$\cap$ is commutative.

$\cap$ is associative.

For a, b in B, if $a \cap b' = c \cap c'$ for some c in B, then $a \cap b = a$.

For a, b in B, if $a \cap b = a$, then $a \cap b' = c \cap c'$ for all c in B.

The first two properties are axioms, and the last two follow from the facts that for all c in B, $c \cap c' = 0$, and $a \cap b' = 0$ iff $a \cap b = a$. We shall prove next that a triple $\langle B, \cap, ' \rangle$ having the properties mentioned above (a precise description appears in the next theorem) may be taken as a formulation of the theory of Boolean algebras. That is, the primitive terms of the initial formulation of the theory can be defined and the axioms (i)–(v) can be derived as theorems.

THEOREM 4.3.1. The following is a formulation of the theory of Boolean algebras. The primitive terms are an unspecified set B, a binary operation $\cap$ in B, and a unary operation $'$ in B. The axioms are as follows.

B_1. $\cap$ is a commutative operation.

B_2. $\cap$ is an associative operation.

B_3. For all a, b in B, if $a \cap b' = c \cap c'$ for some c in B, then $a \cap b = a$.

B_4. For all a, b in B, if $a \cap b = a$, then $a \cap b' = c \cap c'$ for all c in B.*

B_5. The set B has at least two members.

Proof. It remains to prove that the primitive terms of the original formulation can be defined and the axioms derived from a triple $\langle B, \cap, ' \rangle$ satisfying B_1–B_5. As the undefined set and the meet operation of the original formulation, we take B and $\cap$, respectively. A join operation is defined below. The first ten results below (T1–T10)

---

* Axioms B_3 and B_4 are equivalent to the statement "For all a, b, and c in B, $a \cap b' = c \cap c'$ iff $a \cap b = a$." This is easier to remember.

about $\langle B, \cap, ' \rangle$, together with B_1 and B_2, establish the validity of all axioms of the original formulation except the distributive laws. The remainder of the proof is concerned with them. A telegraphic style of presentation is used for ease in reading.

T1. $a \cap a = a$
  Pr. $a \cap a' = a \cap a'$. Now apply B_3.
T2. $a \cap a' = b \cap b'$.
  Pr.   T1 and B_4.

This result justifies the following definition.

D1. $0 = a \cap a'$ and $1 = 0'$.
T3. $a \cap 0 = 0$.
  Pr. $a \cap 0 = a \cap (a \cap a')$,   by D_1;
             $= (a \cap a) \cap a'$,   by B_2;
             $= 0$,   by T1 and D1.
T_4. $a'' = a$.
  Pr.   1. $a'' \cap a' = 0$,   from D1 and B_1.
         2. $a'' \cap a = a''$,   from 1 by B_3.
         3. $a'''' \cap a'' = a''''$,   from 2.
         4. $a'''' \cap a = a''''$,   from 2 and 3 by B_2.
         5. $a'''' \cap a' = 0$,   from 4, by B_4 and D1.
         6. $a' \cap a''' = a'$,   from 5, by B_1 and B_3.
         7. $a''' \cap a' = a'''$,   from 2.
         8. $a''' = a'$,   from 6 and 7.
         9. $a \cap a''' = 0$,   from 8 and D1.
        10. $a \cap a'' = a$,   from 9 by B_3.
        11. $a'' = a$,   from 2 and 10, by B_1.
T5. $a \cap 1 = a$.
  Pr. $a \cap (a \cap a')'' = 0$,   by T4, T1, and D1.
         $a \cap (a \cap a')' = a$,   from the above, by B_3.
         $a \cap 1 = a$,   by D1.
T6. $0 \neq 1$.
  Pr.   1. Assume $0 = 1$.
         2. $a \cap 0 = a$,   from 1 and T5.
         3. $a \cap 0 = 0$,   by T3.
         4. $a = 0$,   from 2 and 3.
         5. This contradicts B_5.
D2. $a \cup b = (a' \cap b')'$.
T7. $(a \cup b)' = a' \cap b'$ and $(a \cap b)' = a' \cup b'$.

Pr.	Both follow from D2 and T4.
T8.	$a \cup b = b \cup a$ and $a \cup (b \cup c) = (a \cup b) \cup c$.
Pr.	The first follows from B_2, and the second follows from B_3 and T4.
T9.	$a \cup a' = 1$.
Pr.	This follows from D2, T4, B_1, and D1.
T10.	$a \cup 0 = a$.
Pr.	This follows from D2, D1, and T4.
T11.	$a \cap (a \cup b) = a$.

Pr.
1. $b' \cap (a \cap a') = 0$,   by T3 and D1.
2. $a \cap (a' \cap b') = 0$,   from 1, by B_1 and B_2.
3. $a \cap (a' \cap b')'' = 0$,   from 2, by T4.
4. $a \cap (a' \cap b')' = a$,   from 3, by B_3.
5. $a \cap (a \cup b) = a$,   from 4, by D2.

T12. $a \cap (a \cap b)' = a \cap b'$.

Pr.
1. $a \cap b'' \cap (a \cap b)' = 0$,   by D1 and T4.
2. $a \cap (a \cap b)' \cap b'' = 0$,   from 1, by B_1.
3. $a \cap (a \cap b)' \cap b'$
   $= a \cap (a \cap b)'$ from 2, by B_3.
4. $a \cap b' \cap (a \cap b)'$
   $= a \cap (a \cap b)'$,   from 3, by B_1.
5. $a \cap b' \cap (a \cap b)'$
   $= a \cap b' \cap (b' \cup a')$,   by T7 and B_1.
6. $a \cap b' \cap (b' \cup a') = a \cap b'$,   by T11.
7. $a \cap (a \cap b)' = a \cap b'$,   from 4, 5, and 6.

T13. $a \cap c = a$, $a \cap c' = 0$ and $a \cup c = c$ are equivalent properties.

Pr.   Left as an exercise.

T14. $a \cap c = a$ and $b \cap c = b$ imply $(a \cup b) \cap c = a \cup b$.

Pr.   Assume that $a \cap c = a$ and $b \cap c = b$. Then $a \cup c = c$ and $b \cup c = c$, by T13. By T11,

$$(a \cup b) \cap [(a \cup b) \cup c] = a \cup b.$$

Two substitutions within the brackets give the desired result.

T15. $a \cap (b \cup c) = (a \cap b) \cup (a \cap c)$ and $a \cup (b \cap c) = (a \cup b) \cap (a \cup c)$.

Pr.
1. $(a \cap b) \cap [a \cap (b \cup c)]$
   $= a \cap b \cap (b \cup c) = a \cap b$,   by B_2, T1, and T11.
2. $(a \cap c) \cap [a \cap (b \cup c)] = a \cap c$, similarly.

3. $[(a \cap b) \cup (a \cap c)] \cap [a \cap (b \cup c)]$
   $= [(a \cap b) \cup (a \cap c)]$,   from 1, 2, and T14.

4. $[a \cap (b \cup c)] \cap [(a \cap b) \cup (a \cup c)]'$
   $= a \cap (b \cup c) \cap (a \cap b)' \cap (a \cap c)'$, by T7.
   $= a \cap b' \cap c' \cap (b \cup c)$,   by B_1 and T12.
   $= a \cap (b \cup c)' \cap (b \cup c)$,
   $= 0$.

5. $[a \cap (b \cup c)] \cap [(a \cap b) \cup (a \cap c)]$
   $= a \cap (b \cup c)$,   from 4 by T13.

6. $a \cap (b \cup c)$
   $= (a \cap b) \cup (a \cap c)$,   from 3 and 5 by B_1.

The proof of the other distributive law is left as an exercise.

The set of axioms in the new formulation of the theory of Boolean algebras is independent. A proof of this requires the determination of a structure $\langle B, \cap, ' \rangle_i$, which satisfies all the axioms except B_i, $i = 1, 2, 3, 4, 5$. Below are defined five systems which demonstrate the independence of the axiom with the corresponding label.

(B_1) $B = \{a,b,c\}$ (B_2) $B = \{a,b,c\}$ (B_3) $B = \{a,b\}$

$\cap$	a	b	c
a	a	a	a
b	a	b	b
c	a	c	c

$'$	
a	b
b	a
c	a

$\cap$	a	b	c
a	a	c	b
b	c	b	a
c	b	a	c

$'$	
a	a
b	c
c	b

$\cap$	a	b
a	a	b
b	b	b

$'$	
a	b
b	b

(B_4) $B = \{A \in \mathcal{P}(\mathbb{Z}^+) | \mathbb{Z}^+ - A \text{ is a finite set}\}$
         $\cap$ is set intersection

$'$ is defined as follows. We note that for each A in B there exists a least positive integer a such that $[a]$, the set of all integers $x \geq a$, is included in A. Then A is the disjoint union of $[a]$ and A_0, a subset of $\{1, 2, \ldots, a - 2\}$ (unless $A = \mathbb{Z}^+$, in which case $A = [1]$). Now we define A' to be $A_0' \cup [a + 1]$, where A_0' is the complement of A_0 in $\{1, 2, \ldots, a - 1\}$ (unless $A = \mathbb{Z}^+$, in which case $A' = [2]$).

(B₅) $$B = \{a\},$$
$$a \cap a = a,$$
$$a' = a.$$

The structure which establishes the independence of B_4 may require some study before it is understood. Some hints that should assist in its analysis are given in Exercise 4.3.2.

Example

4.3.1. Let $B = \{1,2,3,5,6,10,15,30\}$, the set of divisors of 30. For a and b in B, define $a \cap b = \text{lcm } \{a,b\}$ (that is, the least common multiple of a and b) and $a' = 30/a$. It is an easy matter to verify that $\langle B,\cap,'\rangle$ is a Boolean algebra. The corresponding partial ordering of B is this: $a \leq b$ iff a is a multiple of b. Thus, 30 is the least (and zero) element, and 1 is the greatest (and unit) element of the algebra. This ordering of B was used in an example in §1.11 prior to the definition of isomorphism for partially ordered sets. It is left as an exercise to verify that $a \cup b$, which in our second formulation of a Boolean algebra is defined as $(a' \cap b')'$, is gcd $\{a,b\}$, the greatest common divisor of a and b. Thus, if at the outset we had introduced in B, along with the operation $\cap$, a second binary operation $\cup$ by defining $a \cup b$ as the greatest common divisor of a and b, the outcome would have been the same. However, in the process, we would have had to verify the distributive laws, which, in this case, is not a particularly simple matter.

Notice, finally, that the dual of this Boolean algebra is obtained by interchanging "lcm" and "gcd," and "is a multiple of" and "is a divisor of."

Exercises

4.3.1. Prove T13 and the remaining distributive law in the proof of Theorem 4.3.1.

4.3.2. Regarding the system $\langle B,\cap,'\rangle$, which, it is asserted, establishes the independence of B_4, it is clear that B_1 and B_2 hold. Prove that the system satisfies B_3 but not B_4. Hint: for B_3, show that if $C = C_0 \cup [c]$, then $C \cap C' = [c+1]$, and, if $A = A_0 \cup [a]$ and $B = B_0 \cup [b]$, then

$$A \cap B' = \begin{cases} (A \cap B_0') \cup [b+1] & \text{if } a \leq b \\ (A_0 \cap B') \cup [a] & \text{if } a > b \end{cases}$$

4.3.3. Show that each of B_6, B_7, . . . , B_{11} defined below implies B_4 in the presence of B_1, B_2, B_3, and B_4. Infer that each of B_6, B_7, . . . , B_{11} together with B_1, B_2, B_3, and B_5 yields a formulation of the theory of Boolean algebras. For some calculations it is convenient to use the fact that if $\langle B,\cap,'\rangle$ satisfies B_1, B_2, and B_3, then $\langle B,\leq\rangle$, where $a \leq b$ means $a \cap b = a$, is a partially ordered set. So prove this first.

B_6. For all a and b, $a \cap a' = b \cap b'$.

B_7. For all a, $a'' = a$.

B_8. There exists in B an element m such that whenever $x \cap m = x$, $x = m$.

B_9. There exists an integer $n > 1$ such that for all a, the nth iteration of a under $'$ is equal to a.

B_{10}. For all a and b, $a \leq b$ implies $b' \leq a'$.

B_{11}. B is finite.

4.3.4. Verify that the structure defined in Example 4.3.1 is a Boolean algebra. Show that in this algebra $a \cup b = \gcd \{a,b\}$.

4.3.5. Show that the set of divisors of any square-free integer determines a Boolean algebra in exactly the same way as do the divisors of 30.

§4.4. Congruence Relations

We turn to an examination of an aspect of the two given sets of axioms for a Boolean algebra that has not been touched on. It is sufficient to consider the second set of axioms, since the reader will readily see what alterations are required for our remarks to apply to the first set. When the statements labeled B_1, B_2, B_3, B_4, and B_5 were introduced, no mention was made of the precise meaning to be assigned the relation symbolized by "$=$"; rather, it was intended that the reader supply his own version of equality. Suppressing any preconceived notions that we might have in this connection, let us determine a set of conditions which are adequate for our purposes. An analysis of the proofs of T1–T15 in the proof of Theorem 4.3.1 reveals that the following is a sufficient set of conditions.

(E) "$=$" is an equivalence relation.

(S) Let F be an element of the Boolean algebra $\langle B, \cap, ' \rangle$ resulting from elements a, b, ..., of B using the operations in B, and let $a = a_1$, $b = b_1$, ... Then, if F_1 is an element which results from F by the replacement of some or all occurrences of a by a_1, b by b_1, ..., then $F = F_1$.

Now (S) can be derived from the following two simple instances of this substitution principle.

(C) If $a = b$, then $a \cap c = b \cap c$ for all c.
   If $a = b$, then $a' = b'$.

The proof, which we forego, is by induction on the number of symbols in the element F. Thus (E) and (C) insure (E) and (S), and hence (E) together with (C), which are clearly necessary properties, are also sufficient for our purposes. As such, equality is an instance of a congruence relation for a Boolean algebra, a concept we shall discuss after offering an introductory account of congruence relations in a general setting.

Let S be a set and ρ be an equivalence relation on S. If f is a unary operation in S, we shall say that ρ is a congruence relation with respect to f if whenever $a \rho b$ we have $f(a) \rho f(b)$. If g is a binary operation in S, we shall say that ρ is a congruence relation with respect to g if whenever $a \rho c$ and $b \rho d$, we have $g(a,b) \rho g(c,d)$. If σ is a binary relation in S, we shall say that ρ is a congruence relation with respect to σ if whenever $a \rho c$ and $b \rho d$ and $a \sigma b$, we have $c \sigma d$. It is clear how to extend such definitions to n-ary operations and relations in S. Assuming this has been done, we make the following general definition. If $\langle D, . . .\rangle$ is a structure, then an equivalence relation ρ on D will be called a **congruence relation on** $\langle D, . . .\rangle$ iff ρ is a congruence relation with respect to each member operation and relation of the structure. An immediate example of a congruence relation on a structure is the equality relation that is either explicitly or implicitly given for elements of its domain—that equality is a congruence relation is an assumption about the equality relation. At some point in the study of a structure $\langle D, . . .\rangle$, there may be compelling reasons for identifying certain unequal elements of D. This can be achieved by introducing an appropriate equivalence relation ρ on D and directing one's attention to ρ-equivalence classes, that is, elements of D/ρ. If ρ is not merely an equivalence relation but a congruence relation on the structure, then faithful analogues of whatever operations and relations accompany the structure can be introduced in D/ρ. This assertion will gain acceptance from the instances of it that we shall consider in detail.

The first is the study of statement calculi in the light of the theory of Boolean algebras. As background for our presentation, we require only the intuitive discussion of the statement calculus at the beginning of Chapter 2. Let S_0 be a (nonempty) collection of statements. Suppose this set is extended to the smallest set S of statements such that the negation of each member of S is a member of S and each of the conjunction, disjunction, conditional, and biconditional of any two members of S is a member of S. We express this property of S by saying that it is closed with respect to the sentential connectives. A set of statements having these closure properties may be called a statement calculus. Since it was observed that the disjunction, conditional, and biconditional of two statements can be defined in terms of negation and conjunction, we may and shall assume that S is simply the closure of S_0 with respect to these connectives. Then $\wedge$ takes on the role of a binary operation in S and $'$ (which we shall use as the symbol for negation) that of a unary operation in S.

Although the structure $\langle S, \wedge, '\rangle$ is not a Boolean algebra (for instance, if A, B are in S, then $A \wedge B$ is not equal to $B \wedge A$), its properties suggest that there is one lurking in the background. The clue as to how to uncover it lies in the fact that, although two such statements as $A \wedge B$ and $B \wedge A$ are unequal, they are indistinguishable with respect to logical equivalence, that is, $A \wedge B$ eq $B \wedge A$. Now eq is a congruence relation on $\langle S, \wedge, '\rangle$. Indeed, we already know that it is an equivalence relation on S and, using truth tables, it is easily shown that A eq C and B eq D imply that $(A \wedge B)$ eq $(C \wedge D)$ and A' eq C'. The translation in S/eq of the first implication is:

if $[A] = [C]$ and $[B] = [D]$, then $[A \wedge B] = [C \wedge D]$.

Thus, the relation

$$\{\langle\langle[A], [B]\rangle, [A \wedge B]\rangle | [A], [B] \in S/\text{eq}\}$$

is a function on $(S/\text{eq})^2$ into S/eq, that is, a binary operation in S/eq. We shall denote this operation in S/eq by $\wedge$ and its value at $\langle[A],[B]\rangle$ by $[A] \wedge [B]$. So, by definition,

$$[A] \wedge [B] = [A \wedge B].$$

Similarly, that from A eq C follows A' eq C', supplies a proof that a unary operation is defined in S/eq by setting

$$[A]' = [A'].$$

It is a straightforward exercise to verify that $\langle S/\text{eq}, \wedge, '\rangle$ is a Boolean algebra with zero element $[A \wedge A']$ and unit element $[A \vee A']$ for any statement A; this is known as the **Lindenbaum algebra** of the statement calculus S. Notice that 0 is the eq-class of all contradictions of S, and 1 is the eq-class of all tautologies of S. Further, notice that we have our first example of the earlier assertion that given a congruence relation ρ on a structure $\langle D,. . .\rangle$, faithful analogues of operations and relations in D can be introduced in D/ρ.

Frequently the result which we have obtained is stated as "The statement calculus under the connectives 'and' and 'not' is a Boolean algebra." This is misleading, for two reasons. It ignores the fact that there is a statement calculus for each S_0 (actually, it is only the size of S_0 that matters—two statement calculi whose basic sets are in one-to-one correspondence differ only in verbal foliage) and that the elements of the algebra are eq-classes.

Now we turn our attention to congruence relations on Boolean algebras. Let $\langle B, \cap, ' \rangle$ be a Boolean algebra and let θ be a **congruence relation** on it, by which we mean that θ is an equivalence relation on B such that the following hold.

(C_1) If $a \, \theta \, b$, then $a \cap c \, \theta \, b \cap c$ for all c.

(C_2) If $a \, \theta \, b$, then $a' \, \theta \, b'$.

We shall be concerned solely with **proper** congruence relations, that is, those congruence relations different from the universal relation on B. This definition of congruence qualifies as an instance of our general definition because from (C_1) and properties of a Boolean algebra we can deduce

(C_3) If $a \, \theta \, c$ and $b \, \theta \, d$, then $a \cap b \, \theta \, c \cap d$.

For proof, assume that $a \, \theta \, c$ and $b \, \theta \, d$. Then

$$a \cap b \, \theta \, c \cap b \text{ and } b \cap c \, \theta \, d \cap c,$$

by (C_1). Since the meet operation is commutative and θ is transitive, the result follows. The derivation of the dual of (C_3) is left as an exercise. If B/θ is the set of θ-equivalence classes $\bar{a}$, then in B/θ the foregoing result (C_3) becomes the following.

If $\bar{a} = \bar{c}$ and $\bar{b} = \bar{d}$, then $\overline{a \cap b} = \overline{c \cap d}$.

This means that the relation

$$\{\langle \langle \bar{a}, \bar{b} \rangle, \overline{a \cap b} \rangle | \bar{a} \in B/\theta \text{ and } \bar{b} \in B/\theta\}$$

is a function on $(B/\theta) \times (B/\theta)$ into B/θ, that is, an operation in B/θ. We shall denote this operation in B/θ by $\cap$ and its value at $\langle \bar{a}, \bar{b} \rangle$ by $\bar{a} \cap \bar{b}$. So, by definition,

$$\bar{a} \cap \bar{b} = \overline{a \cap b}.$$

Next, it follows directly from (C_2) that if $\bar{a} = \bar{b}$, then $\overline{a'} = \overline{b'}$. Hence, the relation $\{\langle \bar{a}, \overline{a'} \rangle' \bar{a} \in B/\theta\}$ is a function on B/θ into B/θ. We denote this function by $'$ and its value at $\bar{a}$ by $\bar{a}'$. So, by definition,

$$\bar{a}' = \overline{a'}.$$

It is a straightforward exercise to verify that $\langle B/\theta, \cap, ' \rangle$ is a Boolean algebra. For example, to verify B₃, assume that $\bar{a} \cap \bar{b}' = \bar{c} \cap \bar{c}'$. Then, in turn,

$$\bar{a} \cap \bar{b}' = \bar{c} \cap \bar{c}', \qquad \text{by definition of } \bar{x}';$$
$$\overline{a \cap b'} = \overline{c \cap c'}, \qquad \text{by definition of } \bar{x} \cap \bar{y};$$

$$a \cap b' \, \theta \, c \cap c', \qquad\qquad x \, \theta \, y \text{ iff } \bar{x} = \bar{y};$$
$$(a \cap b')' \, \theta \, (c \cap c')', \qquad \text{by } (C_2);$$
$$a' \cup b \, \theta \, 1, \qquad\qquad \text{by property of } \langle B, \cap, ' \rangle;$$
$$(a' \cup b) \cap a \, \theta \, 1 \cap a, \qquad \text{by } (C_1);$$
$$a \cap b \, \theta \, a, \qquad\qquad \text{by property of } \langle B, \cap, ' \rangle;$$
$$\overline{a \cap b} = \bar{a}, \qquad\qquad \bar{x} = \bar{y} \text{ iff } x \, \theta \, y;$$
$$\bar{a} \cap \bar{b} = \bar{a}, \qquad\qquad \text{by definition of } \bar{x} \cap \bar{y}.$$

In summary, we have shown that from a Boolean algebra $\langle B, \cap, ' \rangle$ and a congruence relation θ on it one may derive a Boolean algebra $\langle B/\theta, \cap, ' \rangle$ whose elements are θ-equivalence classes and whose operations are defined in terms of those of the original algebra using representatives of equivalence classes. If θ is different from the equality relation in B, then the derived algebra may be essentially different from the parent algebra. This is true in the first of the following examples.

Examples

4.4.1. Consider the Boolean algebra $\langle \mathcal{P}(\mathbb{Z}), \cap, ' \rangle$ whose elements are the subsets of $\mathbb{Z}$, the set of integers.* We recall the definition of the symmetric difference, $A + B$, of two sets as the set of all objects which are in one of A and B but not both. For A and B in $\mathcal{P}(\mathbb{Z})$ let us define $A \, \theta \, B$ to mean that $A + B$ has a finite number of elements. It is easily verified that θ is an equivalence relation on $\mathcal{P}(\mathbb{Z})$. Further, if $A \, \theta \, B$, then $A \cap C \, \theta \, B \cap C$, since, for all A, B, and C,

$$(A \cap C) + (B \cap C) = (A + B) \cap C,$$

and, hence, if $A + B$ is finite, then so is $(A \cap C) + (B \cap C)$. Finally, if $A \, \theta \, B$, then $A' \, \theta \, B'$, since $A + B = A' + B'$. Thus, θ is a proper congruence relation on the given algebra, and a new Boolean algebra whose elements are θ-equivalence classes results on defining

$$\bar{A} \cap \bar{B} = \overline{A \cap B} \quad \text{and} \quad \bar{A}' = \overline{A'}.$$

That a substantial collapse of elements has taken place on transition from the first to the second algebra is indicated by the fact that, in the first the zero element is $\varnothing$, whereas in the second the zero element, $\bar{\varnothing}$, is the collection of all finite subsets of $\mathbb{Z}$.

4.4.2. The symmetric difference operation used in the preceding example can be defined in any Boolean algebra. By the **symmetric difference** of elements x and y of a Boolean algebra, symbolized $x + y$, we understand the element

$$(x \cap y') \cup (x' \cap y).$$

* We prefer to use prime symbols to denote the operation of complementation relative to $\mathbb{Z}$ in this example, so the bar symbol will be available to denote equivalence classes.

It is an easy exercise to prove that this operation is commutative, idempotent (that is, $x + x = x$), and associative. Other important properties which we shall need later are

$$x + 0 = x,$$
$$(x + y) \cap z = (x \cap z) + (y \cap z),$$
$$x' + y' = x + y.$$

Further, since the symmetric difference is defined in terms of union, intersection, and complementation, if θ is a congruence relation on a Boolean algebra, then $x \theta y$ implies that $x + z \theta y + z$.

Exercises

4.4.1. Prove the dual of property (C_3) for a congruence relation θ, namely,
$(C_3)'$ If $a \theta c$ and $b \theta d$, then $a \cup b \theta c \cup d$.

4.4.2. Complete the proof of the assertion in the text that $\langle B/\theta, \cap, ' \rangle$ is a Boolean algebra if $\langle B, \cap, ' \rangle$ is a Boolean algebra and θ is a proper congruence relation on B.

4.4.3. Prove that the symmetric difference definition has the properties stated in Example 4.4.2.

4.4.4. In §4.6 an atom of Boolean algebra is defined to be a nonzero element a such that if $b \le a$, then either $b = 0$ or $b = a$. Show that there are no atoms in the Boolean algebra of equivalence classes defined in Example 4.4.1.

4.4.5. Referring again to Example 4.4.1, let $A \theta_1 B$ mean that $A \theta B$ and that 3 is not a member of $A + B$. Prove that θ_1 is a congruence relation on $\mathcal{P}(\mathbb{Z})$. Determine the atoms of $\mathcal{P}(\mathbb{Z})/\theta_1$.

§4.5. Ideals

At this point it becomes desirable to simplify our notation by identifying an algebra simply by its basic set. Thus, we shall use the phrase "the Boolean algebra B" in place of "the Boolean algebra $\langle B, \cap, ' \rangle$." Let us consider now the relationship of a Boolean algebra B/θ to the algebra B from which B/θ is derived using a proper congruence relation. Let p be the natural mapping (see §1.9) on the *set* B onto the *set* B/θ, that is, the mapping

$$p: B \longrightarrow B/\theta, \text{ where } p(b) = \bar{b}.$$

Since $\bar{a} \cap \bar{b} = \overline{a \cap b}$ and $\bar{a}' = \overline{a'}$,

$$p(a \cap b) = p(a) \cap p(b) \quad \text{and} \quad p(a') = (p(a))'.$$

That is, p is a "many-to-one" mapping (unless θ is the equality relation on B) which preserves operations. A mapping g on one Boolean algebra,

B, onto another, C, which takes meets into meets and complements into complements, that is,

$$g(a \cap b) = g(a) \cap g(b),$$
$$g(a') = (g(a))',$$

is called a **homomorphism** of B onto C, and C is called a **homomorphic image** of B. If, in addition, g is one-to-one, then g is called an **isomorphism** of B onto C.* If g is an isomorphism of B onto C, then g^{-1} (which exists) is easily proved to be an isomorphism of C onto B, and each algebra is called an **isomorphic image** of the other and each is said to be **isomorphic** to the other. Returning to the case at hand, we may say that p is a homomorphism and B/θ is a homomorphic image of B. That is, each proper congruence relation on a Boolean algebra determines a homomorphic image. Conversely, each homomorphic image C of a Boolean algebra B determines a proper congruence relation on B. Indeed, if $f: B \longrightarrow C$ is a homomorphism, then the relation θ defined by $a\,\theta\,b$ iff $f(a) = f(b)$ is a proper congruence relation on B. The proof is left as an exercise. We continue by showing that B/θ, the algebra of θ-equivalence classes, is isomorphic to C. For this we introduce the relation g, which is defined to be

$$\{\langle \bar{x}, f(x)\rangle | \bar{x} \in B/\theta\}.$$

It is easily seen that g is a function which maps B/θ onto C in a one-to-one fashion and that

$$g(\bar{x} \cap \bar{y}) = g(\overline{x \cap y}) = f(x \cap y) = f(x) \cap f(y) = g(\bar{x}) \cap g(\bar{y}),$$
$$g(\bar{x}') = g(\overline{x'}) = f(x') = (f(x))' = (g(\bar{x}))';$$

that is, g is an isomorphism. Moreover, if p is the natural mapping on B onto B/θ, then we observe that for the given homomorphism $f: B \to C$ we have $f = g \circ p$.† The next theorem summarizes our results.

THEOREM 4.5.1. Let B be a Boolean algebra and θ be a proper congruence relation on B. Then the algebra B/θ of θ-equivalence classes is a homomorphic image of B under the natural mapping on B onto B/θ. Conversely, if the algebra C is a homomorphic image of B, then C is isomorphic to some B/θ. Moreover, if $f: B \to C$ is the

---

* This definition of isomorphism views Boolean algebras as structures with two operations. Another is provided in §1.11 when Boolean algebras are regarded as partially ordered sets. When this latter definition is restricted to Boolean algebras, it coincides with the former.

† This decomposition of f is a special case of that derived in Example 1.8.1 for an arbitrary function.

homomorphism at hand, then $f = g \circ p$, where p is the natural mapping on B onto B/θ and g is an isomorphism of B/θ with C.

Clearly isomorphism is an equivalence relation on the class $\mathcal{C}$ of all homomorphic images of a given Boolean algebra B. It follows from the above theorem, using Exercise 1.9.5, that there is a one-to-one correspondence between $\mathcal{C}$ modulo isomorphism and the set of proper congruences on B. We shall say, more simply, that the proper congruences on B are in one-to-one correspondence with its homorphic images to within an isomorphism.

The importance of the role which proper congruence relations play suggests the problem of practical ways to generate them. One way is provided by a distinguished type of subset of a Boolean algebra, which we define next. A nonempty subset I of a Boolean algebra B is called an **ideal** iff

(i) $x \in I$ and $y \in I$ imply $x \cup y \in I$, and
(ii) $x \in I$ and $y \in B$ imply $x \cap y \in I$.

For example, if $a \in B$, then $\{x \in B | x \leq a\}$ is an ideal; this is the **principal ideal** generated by a, symbolized (a). To show that (a) is an ideal, we note that if $x \in (a)$ and $y \in (a)$, then a is an upper bound of $\{x, y\}$ and, consequently, is greater than or equal to $x \cup y$, the least upper bound of x and y (see Theorem 4.2.2). Thus, $x \cup y \in (a)$. Finally, if $x \in (a)$ and $y \in B$, then $x \cap y \leq a$, since $x \leq a$. Two trivial ideals of B, namely, $\{0\}$ and B, are both principal; indeed, $\{0\} = (0)$, and $B = (1)$. The ideal (0) is the **zero ideal**, and the ideal (1) is the **unit ideal** of B. An ideal of B which is different from B is called a **proper ideal**. The relationship between proper ideals of B and proper congruences on B is given in the following theorem.

THEOREM 4.5.2. If θ is a proper congruence relation on a Boolean algebra B, then $I = \{x \in B | x \,\theta\, 0\}$ is a proper ideal of B and $x \,\theta\, y$ iff $x + y \in I$. Conversely, if I is a proper ideal of B, then the relation θ defined by $x \,\theta\, y$ iff $x + y \in I$ is a proper congruence relation on B such that $I = \{x \in B | x \,\theta\, 0\}$. Thus, the proper congruence relations on B are in one-to-one correspondence with the proper ideals of B; each θ corresponds to the ideal I of elements θ-related to 0.

Proof. Let θ be a proper congruence relation on B and let $I = \{x \in B | x \,\theta\, 0\}$. Then $I \subset B$ and, if $x, y \in I$, then, in turn,

$$x \,\theta\, 0, \qquad x' \,\theta\, 1, \qquad x' \cap y' \,\theta\, 1 \cap y', \qquad x' \cap y' \,\theta\, y', \qquad x \cup y \,\theta\, y.$$

The last fact, when combined with $y \, \theta \, 0$, implies that $x \cup y \, \theta \, 0$, which proves that I satisfies the first of the defining conditions for an ideal. Next, let $x \in I$ and $y \in B$. Since $x \, \theta \, 0$ implies $x \cap y \, \theta \, 0$, the second condition is also satisfied, and I is an ideal.

We prove next that $x \, \theta \, y$ iff $x + y \in I$. Let $x + y \in I$; that is, $x + y \, \theta \, 0$. Then $(x + y) + y \, \theta \, 0 + y$, and hence $x \, \theta \, y$ (where we have used properties of the symmetric difference stated in Example 4.4.2). Conversely, $x \, \theta \, y$ implies that $x + y \, \theta \, y + y$; that is, $x + y \, \theta \, 0$.

Turning to the converse of the foregoing, let I be an ideal of B and define θ as stated in the theorem. Then θ is reflexive (since $x + x = 0 \in I$), symmetric (since $x + y = y + x$), and transitive (since the symmetric difference of two elements of I is in I). Further, $x \, \theta \, y$ implies that $x \cap z \, \theta \, y \cap z$, since if $x \, \theta \, y$, then, in turn, $x + y \in I$, $(x + y) \cap z \in I$, and $(x \cap z) + (y \cap z) \in I$. Finally, $x \, \theta \, y$ implies that $x' \, \theta \, y'$, since $x + y = x' + y'$.

To complete the proof of the converse, we must show that $x \, \theta \, 0$ iff $x \in I$. This follows from the identity $x + 0 = x$.

From the two preceding theorems there follows the existence of a one-to-one correspondence between the homomorphic images, to within an isomorphism, of a Boolean algebra and its proper ideals. If C is a homomorphic image of an algebra B, and $f \colon B \to C$ is a homomorphism, then the associated ideal I, which is called the **kernel** of f, is the set of all elements of B which f maps onto the zero element of C. If θ is the congruence relation on B that corresponds to I, it is customary to write "B/I" instead of "B/θ" and call the algebra so designated (an isomorphic image of C) the **quotient algebra** of B modulo I. If f is an isomorphism, then θ is the equality relation on B and I is the zero ideal. Conversely, it is clear that if the kernel of a homomorphism f can be shown to be the zero ideal, then f is an isomorphism. Therefore, *a homomorphism is an isomorphism iff its kernel is the zero ideal.*

We conclude this section with several general remarks about homomorphisms. Since the operations of union and symmetric difference and the ordering relation are expressible in terms of intersection and complementation, it follows that a homomorphism of a Boolean algebra preserves each of the former. Further, the fact that if f is a homomorphism, then $f(a \cap a') = f(a) \cap (f(a))'$, implies that $f(0)$ is the zero element of the image algebra. By a dual argument, $f(1)$ is the unit element of the image algebra.

Exercises

4.5.1. Prove that if g is an isomorphism of the Boolean algebra B onto the Boolean algebra C, then g^{-1} is an isomorphism of C onto B.

4.5.2. Prove the assertion prior to Theorem 4.5.1 that if $f\colon B \to C$ is a homomorphism, then the relation θ in B defined by $a\,\theta\,b$ iff $f(a) = f(b)$ is a proper congruence on B. Further, prove that $f = g \circ p$, where g and p are the mappings defined in the text.

4.5.3. Prove the assertion following Theorem 4.5.1 that the proper congruence relations on a Boolean algebra B are in one-to-one correspondence with the homomorphic images, to within an isomorphism, of B.

§4.6. Representations of Boolean Algebras

The set-theoretical analogue of our second formulation of the theory of Boolean algebras is that of an algebra of sets. Since it was essentially the structure of such a system that motivated the creation of the axiomatic theory under discussion, an obvious representation problem arises: Is every Boolean algebra isomorphic to an algebra of sets? This we can answer in the affirmative.

We shall begin with the case where the set B has a finite number of elements, although our first definition is applicable to any Boolean algebra. An element a of a Boolean algebra is an **atom** iff $a \neq 0$ and $b \leq a$ implies that either $b = 0$ or $b = a$. For x in B let $A(x)$ denote the set of all atoms such that $a \leq x$. We next derive several properties of atoms and of the sets $A(x)$ for the case of an algebra $\langle B, \cap, '\rangle$ such that B is finite.

A_1. If $x \neq 0$, there exists an atom a with $a \leq x$.

Proof. This is a direct consequence of the finiteness assumption. The details are left as an exercise.

A_2. If a is an atom and $x \in B$, then exactly one of $a \leq x$ and $a \cap x = 0$ holds. Alternatively, exactly one of $a \leq x$ and $a \leq x'$ holds.

Proof. Since $a \cap x \leq a$, either $a \cap x = a$ or $a \cap x = 0$. Moreover, both cannot hold, since $a \neq 0$.

A_3. $A(x \cap y) = A(x) \cap A(y)$.

Proof. First we note that $x \cap y$ is the meet of two elements in B, and $A(x) \cap A(y)$ is the set of those elements common to $A(x)$ and $A(y)$. Now, assume that $a \in A(x \cap y)$. Then $a \leq x \cap y$, and hence $a \leq x$

and $a \leq y$. Thus $a \in A(x) \cap A(y)$. Hence $A(x \cap y) \subseteq A(x) \cap A(y)$. Reversing the foregoing steps establishes the reverse inequality, and hence equality.

A_4. $A(x') = A(1) - A(x)$.

Proof.   First we note that $A(1)$ is the set of all atoms of B. Now let $a \in A(x')$. Then, by A_2, it is false that $a \in A(x)$. Hence, $a \in A(1) - A(x)$. Conversely, if $a \in A(1) - A(x)$, then $a \not\subset A(x)$. Hence, by A_2, $a \in A(x')$.

A_5. $A(x) = A(y)$ iff $x = y$.

Proof.   Assume $x \neq y$. Then at least one of $x \leq y$ and $y \leq x$ is false. Suppose that $x \leq y$ is false. Then $x \cap y' \neq 0$, so that by A_1 there exists an atom $a \leq x \cap y'$. By A_3, $a \in A(x)$ and $a \in A(y')$. Thus, $a \in A(x)$ and, by A_4, $a \not\subset A(y)$. Hence, $A(x) \neq A(y)$. The same conclusion follows similarly if it is assumed that $y \leq x$ is false.

A_6.   If $a_1, a_2, \ldots, a_k$ are distinct atoms, $A(a_1 \cup a_2 \cup \cdots \cup a_k) = \{a_1, a_2, \ldots, a_k\}$.

Proof.   Clearly, $\{a_1, a_2, \ldots, a_k\} \subseteq A(a_1 \cup a_2 \cup \cdots \cup a_k)$. For the converse, assume that $a \in A(a_1 \cup a_2 \cup \cdots \cup a_k)$ and $a \neq a_i$, $i = 1$, $2, \ldots, k$. Then, by A_2, $a \cap a_i = 0$, $i = 1, 2, \ldots, k$, and hence $a = a \cap (a_1 \cup a_2 \cup \cdots \cup a_k) = (a \cap a_1) \cup (a \cap a_2) \cup \cdots \cup (a \cap a_k) = 0$, which is impossible.

THEOREM 4.6.1.   Let B be a Boolean algebra of n elements. Then B is isomorphic to the algebra of all subsets of the set of atoms of B. If m is the number of atoms of B, then $n = 2^m$.

Proof.   Let T be the set of m atoms of B. Then the mapping $A: B \longrightarrow \mathcal{P}(T)$ with $x \longrightarrow A(x)$ is one-to-one by A_5 and onto $\mathcal{P}(T)$ by A_6. According to A_3, the image of a meet in B is the meet of the corresponding images in $\mathcal{P}(T)$. According to A_4, the image $A(x')$ of x' is the complement of the image of x, that is, the relative complement of $A(x)$ in T. Thus, A is an isomorphism.

Then $n = 2^m$ follows from the fact established earlier that the power set of a set of m elements has 2^m members.

COROLLARY.   Two Boolean algebras with the same finite number of elements are isomorphic.

The proof is left as an exercise.

Example

4.6.1. We may illustrate Theorem 4.6.1 with the Boolean algebra defined in Example 4.3.1. By virtue of the definition of the ordering relation for that algebra, the atoms are 6, 10, and 15. Consequently, the algebra is isomorphic to the algebra of all subsets of $\{6,10,15\}$. The mapping defined in the proof of Theorem 4.6.1 matches 2 with $\{6,10\}$ and 30 with $\varnothing$, for example. We also note that the atoms of the dual Boolean algebra are 2, 3, and 5.

Before continuing with the representation theory, we urge the reader to pause and reflect on the extent to which Theorem 4.6.1 clarifies the structure of *finite* Boolean algebras (that is, algebras having a finite number of elements). Indeed, it leaves nothing to be desired in the way of a representation theorem. Possibly its definiteness, both with respect to its arithmetical aspect and the inclusion of an explicit recipe for constructing the asserted isomorphism, will be more fully appreciated when the corresponding result for the infinite case is obtained. For this a different approach must be supplied, since there exist Boolean algebras without atoms (see Exercise 4.4.4). In the infinite case the substitute for an atom is a distinguished type of ideal, which we describe next. Let S be the set of all ideals different from B in the Boolean algebra B. Since $\{0\} \in \mathsf{S}$, it is nonempty. Further, the members of S may be characterized as the ideals of B which do not contain 1. As is true of any collection of sets, S is partially ordered by the inclusion relation, and the concept of a maximal element of S is defined. A maximal element of S is a **maximal ideal** of B. The existence of maximal ideals in an infinite Boolean algebra is secured by an application of a set-theoretic principle known as "Zorn's lemma." It is equivalent to the axiom of choice.* To understand the statement of the lemma, the reader may have to reread that part of §1.11 which follows Theorem 1.11.1.

Zorn's lemma. *A partially ordered set, each of whose simply ordered subsets has an upper bound, contains a maximal element.*

THEOREM 4.6.2. Maximal ideals of a Boolean algebra exist. Indeed, there exists a maximal ideal which includes any preassigned ideal different from B.

Proof. We consider the partially ordered set $\langle \mathsf{S}, \subseteq \rangle$ defined above. If $\mathfrak{C}$ is a simply ordered subset of S, then $A = \{x \in B|$ for some I in $\mathfrak{C}$, $x \in I\}$ is clearly an upper bound for $\mathfrak{C}$. It is a straightforward exercise

* A proof may be found in Stoll (1963), Chapter 2.

to verify that A is an ideal. Moreover, $A \in \mathcal{S}$, since 1 appears in no member of $\mathcal{C}$ and, consequently, does not appear in A. Thus, since every chain in $\mathcal{S}$ has an upper bound in $\mathcal{S}$, Zorn's lemma may be applied to conclude the existence of a maximal element. The same argument when applied to $\{I \in \mathcal{S} | I \supseteq J\}$, where J is a given ideal different from B, yields the existence of a maximal element which includes J.

We prove next a sequence of theorems about maximal ideals of a Boolean algebra B which closely parallels that derived earlier for atoms.

M_1. If $x \neq 1$, there exists a maximal ideal P with $P \supseteq (x)$ or, what amounts to the same, $x \in P$.

Proof. This follows directly from the final statement of Theorem 4.6.2, choosing (x) as the given ideal.

M_2. For each maximal ideal P and each element x of B, exactly one of $x \in P$ and $x' \in P$ holds.

Proof. We note first that for no x is $x \in P$ and $x' \in P$, since it would then follow that $1(= x \cup x') \in P$, which is impossible. Now assume that $x \notin P$, and consider the set Q of all elements of B of the form $b \cup p$ with $b \leq x$ and $p \in P$. Then Q is an ideal, since
  (i) $(b_1 \cup p) \cup (b_2 \cup p) = (b_1 \cup b_2) \cup (p_1 \cup p_2) = b_3 \cup p_3$, and
  (ii) if $y \in B$, then $(b \cup p) \cap y = (b \cap y) \cup (p \cap y) = b_1 \cup p_1$.
Also, $P \subset Q$, since, clearly, $P \subseteq Q$ and $x \in Q$, while $x \notin P$. Thus, $Q = B$, since P is maximal. Hence, for some $b \leq x$ and $p \in P$, $b \cup p = 1$. It follows that $x \cup (b \cup p) = x \cup 1$, or $x \cup p = 1$. Then

$$x' = x' \cap 1 = x' \cap (x \cup p) = (x' \cap x) \cup (x' \cap p) = x' \cap p.$$

By the second part of the definition of an ideal it follows that $x' \in P$.

To continue with the derivation of properties of maximal ideals which parallel, in a complementary sort of way, those for atoms, we introduce the analogue of the sets $A(x)$. If $x \in B$, let $M(x)$ be the set of all maximal ideals P such that $x \notin P$ or, what amounts to the same by virtue of M_2, $x' \in P$. The sets $M(x)$ have the following properties.

M_3. $M(x \cap y) = M(x) \cap M(y)$.

Proof. Let $P \in M(x \cap y)$. Then $(x \cap y)' = x' \cup y' \in P$. Since $x' = x' \cap (x' \cup y')$ and $y' = y' \cap (x' \cup y')$, it follows that $x' \in P$ and

$y' \in P$. Hence $P \in M(x)$ and $P \in M(y)$, or $P \in M(x) \cap M(y)$. Since each of these steps is reversible, the asserted equality follows.

$\mathbf{M_4}$. $M(x') = M(1) - M(x)$, where $M(1)$ is the set of all maximal ideals of the algebra.

Proof.   We have $P \in M(x')$ iff $x' \not\subseteq P$ iff $x \in P$ iff $P \in M(1) - M(x)$.

$\mathbf{M_5}$. $M(x) = M(y)$ iff $x = y$.

Proof.   Assume $x \neq y$. Then at least one of $x \leq y$ and $y \leq x$ is false. It is sufficient to consider the consequences of one of these. Let us say $y \leq x$ is false. Then $x \cup y' \neq 1$, so there exists a maximal ideal P such that $x \cup y' \in P$. Now $(x \cup y')' = x' \cap y \not\subseteq P$, and, hence, by $\mathbf{M_3}$, $P \in M(x')$ and $P \in M(y)$, or $P \not\subseteq M(x)$ and $P \in M(y)$. Thus, $M(x) \neq M(y)$.

The promised representation theorem follows easily from $\mathbf{M_1}$–$\mathbf{M_5}$. It is valid for an arbitrary Boolean algebra, but, in view of the more precise result for finite algebras, it is of interest only in the infinite case. The first proof of this result was given in 1936 by the American mathematician Marshall Stone.

THEOREM 4.6.3.   Every Boolean algebra B is isomorphic to an algebra of sets based on the set of all maximal ideals of B.

Proof.   Let $\mathfrak{M}$ denote the collection of all sets of ideals of the form $M(x)$ for some x in B. According to $\mathbf{M_3}$ and $\mathbf{M_4}$, $\mathfrak{M}$ is an algebra of sets. The mapping $M : B \longrightarrow \mathfrak{M}$, where $x \longrightarrow M(x)$, is onto by the definition of $\mathfrak{M}$ and one-to-one by $\mathbf{M_5}$. Finally, in view of $\mathbf{M_3}$ and $\mathbf{M_4}$, M is an isomorphism.

With the representation theorem for the finite case in mind, it is natural to ask whether the above result cannot be sharpened to read, "Every Boolean algebra is isomorphic to the algebra of all subsets of some set." To discuss this matter we make two definitions. A Boolean algebra B is called **atomic** iff for each nonzero element b of B there exists an atom a of B with $a \leq b$. A Boolean algebra B is called **complete** iff for each nonempty subset A of B, lub A exists relative to the standard partial ordering of B. This definition has significance only when A is infinite, since in any Boolean algebra each pair, and consequently each finite set of elements, has a least upper bound. Now it is clear that the algebra of all subsets of a set is both atomic and complete.

It is left as an exercise to prove that each of these properties is preserved under an isomorphism. Hence, an algebra which fails to have either property cannot be isomorphic to an algebra of all subsets of a set. Since, as noted in Exercise 4.4.4, the algebra defined in Example 4.4.1 is not atomic, the question is settled in the negative. The same conclusion is provided by the algebra defined in Exercise 4.2.6, since, as the reader may show, it is not complete.

The above pair of conditions, which are necessary in order that a Boolean algebra be isomorphic to the algebra of all subsets of a set, are also a sufficient set. This is our next theorem.

THEOREM 4.6.4. Necessary and sufficient conditions that a Boolean algebra be isomorphic to the algebra of all subsets of some set are that B be complete and atomic. In this event, B is isomorphic to the algebra of all subsets of its set of atoms.

Proof. Since the necessity of these conditions has already been observed, we turn to a proof of their sufficiency. Suppose, therefore, that B is complete and atomic and let T be the set of all atoms of B. As in the proof of the finite case, let $A(x)$ denote the set of atoms a for which $a \leq x$. Then, exactly as in the finite case, it can be proved that the mapping A on B into $\mathcal{P}(T)$ has properties A_3 and A_4 (now, of course, property A_1 is an assumption). This means that A is a homomorphism on B onto an algebra of subsets of T. If U is an arbitrary subset of T, then, by the assumed completeness, U has a least upper bound, u say, in B. Then $A(u) = U$ (this is a generalization of A_6 for the finite case), so A is onto $\mathcal{P}(T)$.

All that is needed to complete the proof is to show that A is one-to-one—that is, that the kernel of A is the zero ideal. This follows from the atomicity of B; if $x \neq 0$, then $A(x) \neq 0$, so $A(x) = \varnothing$ iff $x = 0$.

We conclude this section with a justification of an assertion made in §1.5 to the effect that, via an axiomatic approach to the algebra of sets, one may conclude that every identity of algebras of sets is derivable from 1 to 5 and $1'$ to $5'$ of Theorem 1.5.1. To begin, we formulate the theory of Boolean algebras as a first-order theory $\mathfrak{B}$. This theory has one predicate letter P_1^2, three function letters f_1^1, f_1^2, and f_2^2, and individual constants a_1 and a_2. For terms s and t we abbreviate $P_1^2(s,t)$ by $s = t$, $f_1^1(s)$ by s', $f_1^2(s,t)$ by $s \cup t$, $f_2^2(s,t)$ by $s \cap t$, and a_1 and a_2 by 0 and 1. The proper axioms are the formal analogues of (i), (i)$'$, $\dots$, (v) in §4.1 (note that each is a closed wf) plus

$(x_1)(x_1 = x_1),$

$(x_1)(x_2)(x_1 = x_2 \rightarrow x_2 = x_1),$

$(x_1)(x_2)(x_3)(x_1 = x_2 \wedge x_2 = x_3 \rightarrow x_1 = x_3),$

$(x_1)(x_2)(x_1 = x_2 \rightarrow x_1' = x_2'),$

$(x_1)(x_2)(x_3)(x_1 = x_2 \rightarrow x_1 \cap x_3 = x_2 \cap x_3 \wedge x_3 \cap x_1 = x_3 \cap x_2),$

$(x_1)(x_2)(x_3)(x_1 = x_2 \rightarrow x_1 \cup x_3 = x_2 \cup x_3 \wedge x_3 \cup x_1 = x_3 \cup x_2).$

Consider now the statement

(S) Every wf of $\mathfrak{B}$ which is true in every model of $\mathfrak{B}$ (that is, in every model of the set X of proper axioms of $\mathfrak{B}$) is a theorem of $\mathfrak{B}$.

Assuming that (S) is a theorem, we can deduce the correctness of our original assertion. Indeed, let I be an identity for algebras of sets and let I^* be the corresponding wf of $\mathfrak{B}$. Let $\mathfrak{M}$ be a model of $\mathfrak{B}$. According to the representation theorem for Boolean algebras, there is an algebra of sets, $\mathfrak{C}$, isomorphic to M. Then I and, hence, I^* are true in $\mathfrak{C}$. Since $\mathfrak{M}$ is isomorphic to $\mathfrak{C}$, it follows that I^* is true in $\mathfrak{M}$. From (S) we may conclude that I^* is a theorem of $\mathfrak{B}$, but this means that I^* is deducible from (i), (i)$'$, . . . , (v). Thus I is deducible from (1) to (5) and (1)$'$ to (5)$'$ of Theorem 1.5.1. Observe, in addition, that "is deducible from" now has a precise meaning.

The proof of our initial assertion thus becomes complete when (S) is proved. For this we observe first that (S) is equivalent to

(S)$'$ Every logical consequence of the set X of proper axioms of $\mathfrak{B}$ is a theorem of $\mathfrak{B}$.

(Indeed, the very definition of logical consequence implies that if a wf A of $\mathfrak{B}$ is a logical consequence of X, then A is true in every model of X; since every member of X is closed, an observation following the Corollary to Theorem 2.8.1 establishes the converse.) But the validity of (S)$'$ follows immediately from the first result stated in the paragraph below the proof of Theorem 3.6.2.

Exercises

4.6.1. Prove property A_1 of atoms in a finite Boolean algebra.

4.6.2. Prove the Corollary to Theorem 4.6.1.

4.6.3. (a) Prove the converse of property M_2 of maximal ideals to obtain a characterization of maximal ideals among the set of proper ideals.

 (b) Prove that maximal ideals can also be characterized as those ideals I of a Boolean algebra B such that B/I has just two elements.

4.6.4. (a) In Exercise 4.2.9, there is defined the Boolean algebra $\mathfrak{C}$ of all subsets A of $\mathbb{Z}^+$ such that either A or $\overline{A}$ is finite. Prove that the collection $\mathfrak{C}$ of all finite subsets of $\mathbb{Z}^+$ is a maximal ideal of $\mathfrak{C}$.

(b) The same collection $\mathfrak{C}$ is an ideal of the algebra $\mathcal{P}(\mathbb{Z}^+)$. Prove that $\mathfrak{C}$ is not a maximal ideal of this algebra and determine a maximal ideal which includes $\mathfrak{C}$.

4.6.5. Devise a proof of Theorem 4.6.3 for the case of a *denumerable* Boolean algebra B that does not employ Zorn's lemma. (Hint: Prove by induction that if B is denumerable then there exists a maximal ideal which includes any pre-assigned ideal.)

4.6.6. Prove that the Boolean algebra in Exercise 4.6.3(a) is not complete by showing that the collection of all unit sets of positive even integers has no least upper bound.

4.6.7. Prove that an isomorphic image of a complete Boolean algebra is complete and that an isomorphic image of an atomic algebra is atomic.

4.6.8. Prove that every ideal of a Boolean algebra B is principal iff B is finite. (Note: The proof that B is finite if every ideal is principal is difficult.)

4.6.9. Define a half-open interval to be a subset of $\mathbb{Q}$ which is either empty or of one of the forms

$$\{x \in \mathbb{Q} | x < b\}, \qquad \{x \in \mathbb{Q} | a \leq b < x\}, \qquad \{x \in \mathbb{Q} | a \leq x\},$$

where a and b are in $\mathbb{Q}$. Let B be the collection of all finite set-theoretic unions of half-open intervals.

(a) Show that $\langle B, \subseteq \rangle$ is a Boolean algebra.
(b) Show that this algebra has no atoms.
(c) Assuming as known the fact that $\mathbb{Q}$ is denumerable, show that B is denumerable.
(d) Show that $\langle B, \subseteq \rangle$ is essentially the only atomless Boolean algebra having a denumerable number of elements by proving that any two such algebras are isomorphic.

§4.7. Free Boolean Algebras

In §4.4 we considered an intuitive statement calculus as the closure S with respect to "and" and "not" of an initial set S_0 of statements. Then we formed S/eq, transferred the operations defined for S to S/eq, and proved that the resulting structure is a Boolean algebra. As such we have an illustration of a rather familiar situation in algebra, namely, the introduction of such a congruence relation ρ on a given structure $\langle X, \ldots \rangle$ that the operations and relations transferred from X to X/ρ have not only all properties of the parent operations and relations but additional properties. In what follows we imitate in a purely formal way the construction just described. The congruence we introduce is intended to be the least restrictive so that the resulting quotient system is a Boolean algebra.

Let S_0 be an arbitrary nonempty set and $\wedge$ and $'$ be two symbols which do not designate elements of S_0. We give an inductive definition of a set S whose elements are certain finite sequences of elements of $S_0 \cup \{\wedge, '\}$ together with parentheses.

(I) If $s \in S_0$, then $s \in S$.

(II) If $t \in S$, then $(t)' \in S$.

(III) If $s, t \in S$, then $(s) \wedge (t) \in S$.

(IV) The only members of S are those resulting from a finite number of applications of (I), (II), and (III).

The result is a structure $\langle S, \wedge, ' \rangle$ consisting of a set S (in which we regard two elements as equal iff they are identical), together with a binary and a unary operation in S. We now investigate conditions that a relation on S must satisfy if the associated quotient structure is a Boolean algebra. On the basis of the discussion in §4.4, necessary and sufficient conditions that such a relation θ must satisfy are that it be an equivalence relation different from the universal relation on S (the latter requirement reflects the fact that a Boolean algebra has more than one element) and that the following hold for all elements of S.

If $s \theta t$, then $s \wedge u \theta t \wedge u$ for all u.*

If $s \theta t$, then $s' \theta t'$.

(C) $\quad s \wedge t \theta t \wedge s$.

$\quad s \wedge (t \wedge u) \theta (s \wedge t) \wedge u$.

If $s \wedge t' \theta u \wedge u'$ for some u, then $s \wedge t \theta s$.

If $s \wedge t \theta s$, then $s \wedge t' \theta u \wedge u'$ for all u.

In defense of our assertion we note that the first two parts of (C) are necessary and sufficient conditions that the operations in S induce operations in S/θ in a natural way, and the remaining four parts constitute a minimal set of conditions which insure that the resulting system is a Boolean algebra. Parenthetically, we remark that at times, when an equivalence relation satisfying (C) is introduced into $\langle S, \wedge, ' \rangle$, it is more natural to continue with the elements of S (instead of those of S/θ) as the basic objects. This attitude is reflected in referring to the system $\langle S, \wedge, ' \rangle$ as a Boolean algebra with respect to θ.

There is no question concerning the existence of equivalence relations satisfying (C), since if members of S are interpreted as truth functions, then, as observed in §4.4, the eq relation satisfies (C). We

---

* Here we begin to follow the usual mathematical conventions of omitting superfluous parentheses.

consider now the set $\mathcal{C}$ of all equivalence relations satisfying (C) and let μ denote the intersection of the collection $\mathcal{C}$. It is left as an exercise to prove that $\mu \in \mathcal{C}$ and, consequently, is the smallest member of $\mathcal{C}$ in the sense that it relates the fewest possible pairs of elements of S. The Boolean algebra S/μ is called the **free Boolean algebra** generated by S_0. In this context the word "free" is intended to suggest that the elements of the algebra are as unrestricted as is possible if they are to have the structure of a Boolean algebra. Intuitively this is clear, since the only relations which have been imposed upon them are a necessary and sufficient set to insure that they do have that structure. There are alternative definitions of a free Boolean algebra that are more exotic; our old-fashioned one has the merit that it simultaneously disposes of the existence of such algebras.

It is possible to give an interesting characterization of the congruence relation μ. To this end we consider the μ-equivalence class

$$\mathcal{U} = [(u \wedge u')']_\mu$$

for some u in S. This class is independent of u since it includes all members of S having the same form. This follows from the fact that if $\theta \in \mathcal{C}$, then $(s \wedge s')'\, \theta\, (u \wedge u')'$ for all s in S, and hence $(s \wedge s')'\, \mu\, (u \wedge u')'$ for all s in S. Since the zero element of S/μ is $[u \wedge u']_\mu$, $\mathcal{U}$ is the unit element of S/μ. It is left as an exercise to prove that if $s,\, t \in S$, then

$$s \,\mu\, t \quad \text{iff} \quad (s \wedge t')' \wedge (s' \wedge t)' \in \mathcal{U}$$

or, introducing $s \leftrightarrow t$ as an abbreviation for $(s \wedge t')' \wedge (s' \wedge t)'$,

$$s \,\mu\, t \quad \text{iff} \quad s \leftrightarrow t \in \mathcal{U}.$$

This characterization of μ in terms of $\mathcal{U}$ is opaque until S is interpreted as the set of formulas of a statement calculus. Then it will be recognized that μ is to be interpreted as the eq relation and that $\mathcal{U}$ becomes the set of valid formulas of S. Finally, the characterization of μ in terms of $\mathcal{U}$ is simply the formal version of Theorem 2.3.2 (namely, s eq t iff $\vDash s \leftrightarrow t$).

The same interpretation of S suggests, as an alternative approach to the definition of the free Boolean algebra generated by S_0, the introduction of the set $\mathcal{U}$ first, followed by the *definition* of μ in terms of $\mathcal{U}$. This is possible using some formulation of a statement calculus as an axiomatic theory. The starting point is the inductive definition of the set S in terms of the elements of $S_0 \cup \{\wedge, '\}$, just as before. We now wish to obtain the set $\mathcal{U}$ as that subset of S which, under the interpretation of S, constitutes the tautologies. This is possible using the results of §3.6.

Introducing $s \to t$ as an abbreviation for $(s \wedge t')'$, we define a subset V of S as follows.

    (I) Any member of S that has one of the following three forms is a member of V:

$$s \to (t \to s),$$
$$(u \to (s \to t)) \to ((u \to s) \to (u \to t)),$$
$$(s' \to t') \to (t \to s).$$

   (II) If s and t are members of S such that both s and $s \to t$ are members of V, then t is a member of V.

  (III) An element of S is a member of V iff it can be accounted for using (I) or (II).

The desired conclusion, that $\upsilon = V$, is then secured via the completeness theorem (Theorem 3.5.3) and its converse (Theorem 3.5.4). In terms of V, the relation μ may now be defined by

$$s \, \mu \, t \quad \text{iff} \quad s \leftrightarrow t \in V.$$

Although statement calculi served as our inspiration for introducing the concept of a free Boolean algebra, now that the latter concept has been firmly established, we may turn matters around and describe the Lindenbaum algebra of a statement calculus as simply the free Boolean algebra generated by the set of prime formulas of the calculus in question.

Exercises

4.7.1. Show that the relation μ is a member of $\mathcal{C}$.

4.7.2. Show that $s \, \mu \, t$ iff $s \leftrightarrow t \in \upsilon$.

References

Birkhoff, G. *Lattice theory*. Providence, 3d ed., 1967.

Cantor, G. *Contributions to the founding of the theory of transfinite numbers*. Translated by P. E. B. Jourdain. Chicago, 1915. Reprinted by Dover Publications, Inc.

Cohen, P. *Set theory and the continuum hypothesis*. New York, 1966.

Dedekind, R. *Essays on the theory of numbers*. Translated by W. W. Beman. Chicago, 1901.

Feferman, S. *The number systems*. Reading, Pa., 1964.

Fraenkel, A., and Y. Bar-Hillel. *Foundations of set theory*. Amsterdam, 1958.

Frege, G. *Begriffsschrift, eine der arithmetischen nachebildete Formelsprache des reinen Denkens*. Halle, 1879.

———. *Die Grundlagen der Arithmetik*. Breslau, 1884.

———. *The basic laws of arithmetic*. Translated by M. Furth. Berkeley, 1964.

Hilbert, D. *Grundlagen der Geometrie*. Berlin, 7th ed., 1899.

———. *The foundations of geometry*. Translated by E. J. Townsend. Chicago, 1902.

Kleene, S. C. *Introduction to metamathematics*. Princeton, 1952.

———. *Mathematical logic*. New York, 1967.

Margaris, A. *First-order mathematical logic*. Waltham, 1967.

Mendelson, E. *Introduction to mathematical logic*. Princeton, 1964.

Rubin, H., and J. Rubin. *Equivalents of the axiom of choice*. Amsterdam, 1963.

Stoll, R. R. *Set theory and logic*. San Francisco, 1963.

van Heijenoort, J., ed. *From Frege to Gödel: A sourcebook in mathematical logic, 1879–1931.* Cambridge, Mass., 1967.

Whitehead, A. N., and B. Russell. *Principia mathematica* I–III. Cambridge, Eng., 1910–1913.

Symbols

Symbol	Meaning	Page
$\mathbb{Z}$	set of integers	3
$\mathbb{Q}$	set of rational numbers	3
$\mathbb{R}$	set of real numbers	3
$\mathbb{C}$	set of complex numbers	3
$\mathbb{Z}^+$	set of positive integers	3
$\mathbb{Q}^+$	set of positive rationals	3
$\mathbb{R}^+$	set of positive reals	3
$\in$	member of	5
$\notin$	not member of	5
$=$	equals	5
$\neq$	unequal to	5
$\{x_1, x_2, \ldots, x_n\}$	set whose members are $x_1, x_2, \ldots, x_n$	6
$\{x \mid P(x)\}$	set of all x such that $P(x)$	9
$\{x \in A \mid P(x)\}$	set of all x in A such that $P(x)$	10
$\subseteq, \supseteq$	included in, includes	12
$\subset$	properly included in	12
$\varnothing$	empty set	13
$\mathcal{P}(A)$	power set of A	13
$A \cup B$	union of two sets	15
$A \cap B$	intersection of two sets	15
$\overline{A}$	absolute complement	15
$X - A$	relative complement	16
$A + B$	symmetric difference of two sets	16
$\langle x, y \rangle$	ordered pair	26
$\langle x_1, x_2, \ldots, x_n \rangle$	ordered n-tuple	27
$x \, \rho \, y$	x is ρ-related to y	27

Symbol	*Meaning*	*Page*
D_ρ	domain of relation	28
R_ρ	range of relation	28
F_ρ	field of relation	28
$X \times Y$	cartesian product	29
ι_X	identity relation	29
$\rho[A]$	set of ρ-relatives of A	29
$\rho\|\sigma$	relative product	30
ρ^{-1}	inverse	31
$[x]$	equivalence class	33
$\mathbb{Z}_n$	set of residue classes modulo n	35
X/ρ	quotient set	35
$f[A]$	image of A under f	38
$\lambda x[f(x)]$	lambda notation	39
$f: x \mapsto y$	f maps x onto y	39
$f: X \longrightarrow Y,$ $X \xrightarrow{f} Y$	map on X into Y	40
Y^X	set of all functions on X into Y	40
$f \restriction A$	restriction of f to A	40
i_X	identity map on X	40
2^X	set of all functions on X into $\{0,1\}$	41
χ_A	characteristic function of A	41
X^n	set of n-tuples in X	41
$g \circ f$	composite of functions	43
$\langle f, Y \rangle$	map	45
f^n	nth iterate of a function	47
$\leq$	less than or equal to	52
$<$	less than	52
$\nleq$	not less than or equal to	52
lub A, sup A	least upper bound of A	56
glb A, inf A	greatest lower bound of A	56
$\mathbb{N}$	set of natural numbers	60
$\sim$	not	71
$\wedge$	and	71
$\vee$	or	71
$\rightarrow$	implies	71
$\leftrightarrow$	equivalent to	71
$\top$	truth	73
F	falsity	73

Symbol	*Meaning*	*Page*
$\models A$	validity (statement calculus)	80
A eq B	equivalence (statement calculus)	82
$A_1, A_2, \ldots, A_m \models B$	B is a consequence of $A_1, A_2, \ldots, A_m$ (statement calculus)	91
$(\forall x), (x)$	for all x	108
$(\exists x)$	there exists an x such that	108
p_j^n	predicate letter	114
f_j^n	function letter	114
wf	well-formed formula	114
$(A \wedge B)$	$(\sim(A \rightarrow (\sim B)))$	115
$(A \vee B)$	$((\sim A) \rightarrow B)$	115
$(A \leftrightarrow B)$	$((A \rightarrow B) \wedge (B \rightarrow A))$	115
$(\exists x)A$	$\sim((x)(\sim A))$	115
$\langle D, g \rangle$	interpretation	119
$\models_I A[s]$	I satisfies A with s	121
$s(x_i|d)$	value of $s: V \rightarrow D$ at x_i is d	125
$\models A$	validity (first-order logic)	125
A eq B	equivalence of wfs	125
$A(x_1, x_2, \ldots, x_n)$	a wf possibly having free occurrences of $x_1, x_2, \ldots, x_n$	126
$X \models B$	B is a consequence of a set of wfs	132
$A_1, A_2, \ldots, A_m \models B$	B is a consequence of wfs $A_1, A_2, \ldots, A_m$	132
$\mathfrak{A} = \langle D, \langle R_i \rangle, \langle F_i \rangle, \langle d_i \rangle \rangle$	structure	146
$\sigma = \langle k, l, m, \langle m_i \rangle, \langle n_i \rangle \rangle$	signature	146
$X \vdash_{\mathfrak{T}} A$	A is deducible in $\mathfrak{T}$ from X	164
$B_1, B_2, \ldots, B_n \vdash A$	A is deducible from $B_1, B_2, \ldots, B_n$	164
$\vdash A$	A is a theorem	164
N	first-order arithmetic	171

Index